ACCESS Health Press

Fusion!
The Melding of Human
and Machine Intelligence

~•

William A. Haseltine Ph.D.
And
Griffin McCombs

Recent Books by William A. Haseltine

Affordable Excellence: The Singapore Healthcare Story; William A Haseltine (2013)

Improving the Health of Mother and Child: Solutions from India; Priya Anant, Prabal Vikram Singh, Sofi Bergkvist, William A. Haseltine & Anita George (2014)

Modern Aging: A Practical Guide for Developers, Entrepreneurs, and Startups in the Silver Market; Edited by Sofia Widén, Stephanie Treschow, and William A. Haseltine (2015)

Aging with Dignity: Innovation and Challenge is Sweden-The Voice of Care Professionals; Sofia Widen and William A. Haseltine (2017)

Every Second Counts: Saving Two Million Lives. India's Emergency Response System. The EMRI Story; William A Haseltine (2017)

Voices in Dementia Care; Anna Dirksen and William A Haseltine (2018)

Aging Well; Jean Galiana and William A. Haseltine (2019)

World Class. Adversity, Transformation and Success and NYU Langone Health; William A. Haseltine (2019)

Science as a Superpower: My Lifelong Fight Against Disease And The Heroes Who Made It Possible; William A. Haseltine (2021)

The COVID-19 Textbook: Science, Medicine, and Public Health; William A. Haseltine and Roberto Pataraca (2023)

Living ebooks

A Family Guide to Covid: Questions and Answers for Parents, Grandparents, and Children; William A. Haseltine (2020)

A Covid Back To School Guide: Questions and Answers for Parents and Students; William A. Haseltine (2020)

Covid Commentaries: A Chronicle of a Plague, Volumes I, II, III, IV, V, and VI; William A. Haseltine (2020)

My Lifelong Fight Against Disease: From Polio and AIDS to Covid-19; William A. Haseltine (2020)

Variants!: The Shape-Shifting Challenge of Covid-19 Vaccine Evasion & Reinfection; William A. Haseltine (2021)

Covid Related Post-traumatic Stress Disorder (CV-PTSD): What It Is And What To Do About It; William A. Haseltine (2021)

Natural Immunity And Covid-19: What It Is And How It Can Save Your Life; William A. Haseltine (2022)

Omicron: From Pandemic to Endemic; William A. Haseltine (2022)

Monoclonal Antibodies: The Once and Future Cure for Covid-19; William A. Haseltine and Griffin McCombs (2023)

The Future of Medicine: Healing Yourself: Regenerative Medicine Part One; William A. Haseltine (2023)

Viroids and Virusoids: Nature's Own mRNAs; William A. Haseltine and Koloman Rath (2023)

CAR T: A New Cure for Cancer, Autoimmune and Inherited Disease; William A. Haseltine and Amara Thomas (2023)

Ending Hepatitis C: A Seven-step Plan for a Successful Eradication Program: A Roadmap for Ending Endemic Disease Globally; William A. Haseltine and Kaelyn Varner (2023)

Better Eyesight: What You and Modern Medicine Can Do to Improve Your Vision; William A. Haseltine and Kim Hazel (2024)

Acknowledgments

We thank the ACCESS Health US team, Koloman Rath, Amara Thomas, and Roberto Patarca, for their support in creating this book. An additional special thank you to Amara Thomas for serving as editor for the piece.

This work is supported by ACCESS Health International (www.accessh.org).

~◡

Dedication

William Haseltine, Ph.D.

To my wife, Maria Eugenia Maury; my children Mara and Alexander; my stepdaughters Karina, Manuela, and Camila;

my grandchildren Pedro Agustin, Enrique Mattias, and Carlos Eduardo; and last but not least, our three dogs, Sky, Luna, and Ginger.

Griffin McCombs, MPA

To my family for love and support;

To my friends for their kindness and understanding;

To my cats for keeping me on my toes;

And to all those who have challenged and encouraged me on my journey.

~

Contents

Prologue: The Fusion Question

This book embarks on a journey to explore one of the most profound and urgent questions of our time: Can we truly merge human and machine intelligence? This is not a mere exercise in speculation; it is an inquiry rooted in the groundbreaking scientific and technological advances already reshaping the landscape of human potential. The fusion of brain and machine is no longer a distant dream — it is happening today, primarily through therapeutic applications designed to restore lost functions, such as speech and movement, to those affected by severe injuries or illnesses. Yet, as we will discover in the chapters ahead, the potential of brain-machine interfaces extends far beyond their initial medical uses.

As we delve deeper into this topic, we will explore how these interfaces might soon enable two-way communication — allowing the human brain not only to control external devices with ever-increasing precision, but also enabling machines to send information directly into our minds. Imagine a future where knowledge, languages, and even specialized skills can be downloaded straight into the brain, bypassing the traditional methods of learning. While the challenges in realizing this vision are substantial, the accelerating pace of research suggests that such capabilities are no longer confined to the realm of science fiction, but are within the reach of our lifetimes.

With each passing year, the fields of neuroscience and artificial intelligence (AI) are converging, breaking down the walls between human cognition and machine capabilities. Neuroscience seeks to uncover the mysteries of the human brain, while AI strives to replicate—or even surpass—human intelligence in machines. These once separate, even opposing, domains are now becoming intertwined, paving the way for new possibilities that enhance human abilities, redefine our relationship with technology, and challenge our very understanding of what it means to be human.

The fusion of mind and machine is not just a technological challenge, but a philosophical one. What happens when human consciousness merges with machine intelligence? What new realities will emerge from this fusion, and how will it affect our identities, our relationships, and our society? These are the questions this book seeks to explore, guiding readers through the current state of brain-machine interfaces while projecting forward into their boundless potential.

We begin by examining the current landscape of brain-machine interfaces, highlighting the incredible therapeutic applications already transforming lives. Through the stories of individuals like Gert-Jan Oskam, Matthew Angle, Jens Naumann, and Casey Harrell, we will witness firsthand how these technologies have restored lost functions, enhanced abilities, and opened new avenues for communication for those living with severe disabilities. These personal accounts not only showcase the technological feats achieved but also the profound human impact of fusing mind and machine, offering a powerful reminder of the resilience of the human spirit and the transformative potential of this technology.

From there, we step back to explore the historical context that has brought us to this point. The evolution of brain-machine interfaces is not a new phenomenon, but rather the culmination of centuries of inquiry into the mind-body connection. By tracing this rich history, we gain a deeper appreciation of the progress made and the potential future directions of this field. Understanding where we have come from is essential to navigating where we are headed, especially as the pace of innovation continues to accelerate.

Next, we turn our focus to the technology enabling brain-machine fusion, with particular attention to the role of electrodes—the critical points of contact between the human brain and external machines. We will examine the differences between invasive and non-invasive electrodes, exploring the trade-offs between precision and safety, and highlighting the cutting-edge innovations that promise to make these devices more effective and accessible. The development of electrode technology is not merely a technical endeavor; it is the cornerstone upon which the entire vision of human-machine fusion rests.

As we move forward, we shift our attention to the future applications of brain-machine interfaces, particularly the role of artificial intelligence. AI is not simply a tool for interpreting neural data; it is the driving force that will allow the fusion of human and machine intelligence to become seamless and scalable. We will explore how AI is enhancing the capabilities of brain-machine interfaces, making them more intuitive, responsive, and powerful. Furthermore, we will consider speculative—but plausible—advances, such as direct brain-to-brain communication and the possibility of downloading information directly into the human brain. These innovations could revolutionize learning, creativity, and even our understanding of human identity, presenting extraordinary opportunities alongside profound ethical dilemmas.

Throughout this exploration, it is crucial to confront the ethical, philosophical, and societal implications of merging human and machine intelligence. What happens to privacy, autonomy, and cognitive liberty when our thoughts can be read or influenced by external devices? How will these technologies impact social inequalities—potentially creating new forms of disparity or bridging existing gaps? In the final chapters of this book, we will tackle these questions, inviting readers to engage in informed public discourse about the future of human-machine integration.

Throughout this journey, we aim to maintain a balanced perspective, acknowledging both the incredible possibilities and the significant risks associated with brain-machine interfaces. Our goal is to provide readers with the knowledge and insights to critically evaluate developments in this rapidly evolving field and to actively participate in shaping its ethical and responsible implementation. As you read, we encourage you to envision the future of human-machine fusion and consider how it will shape the next chapter of human evolution. This is not just a technological journey—it is a profoundly human one, with the potential to redefine what it means to think, to learn, and to live in a world where the boundaries between mind and machine are increasingly blurred.

Introduction: Fusing Mind and Machine

❧

Imagine a world where your thoughts effortlessly shape the reality around you—where a mere flicker of intention allows you to command devices, convey ideas without words, and unlock boundless reservoirs of knowledge. This vision of seamless interaction between the human mind and technology is no longer confined to the realm of dreams. It is steadily emerging, fueled by rapid advances in neuroscience, engineering, and artificial intelligence. The merging of the human brain with machines stands as one of the most revolutionary endeavors of our time, poised to redefine the essence of what it means to be human.

At its core, the aspiration to merge consciousness with technology reflects a timeless yearning to surpass the inherent limitations of our bodies and minds. Throughout the ages, humanity has been captivated by the possibility of transcending the ordinary, of achieving abilities that elevate us beyond flesh and bone. Ancient myths spoke of heroes gifted with extraordinary powers by divine forces or enchanted relics. These tales blurred the lines between the mortal and the divine, hinting at a future where the extraordinary might become attainable.

As centuries passed and science supplanted myth, this enduring fascination found new expression. The modern era gave rise to visions of human-machine integration, with science fiction imagining futures where biology and technology harmonize. These speculative narratives, though fantastical, were rooted in an expanding understanding of scientific potential. They offered glimpses of a reality where the boundaries separating humanity and machines could be not merely crossed but entirely erased.

What once resided in the realm of imagination is now taking form in laboratories and clinics worldwide. The fusion of mind and machine is no longer theoretical; it has become a rapidly evolving field, reshaping how we perceive health, communication, and even the fabric of consciousness itself. The implications are as vast as they are profound, promising to touch every facet of human existence with transformative possibilities.

~⁀

The journey toward merging the human brain with machines began with a simple yet profound desire: to uncover the mysteries of the mind. Early neuroscientists were akin to daring explorers, navigating the uncharted wilderness of human cognition. Armed with curiosity and rudimentary tools, they sought to map the brain's labyrinthine structure, uncovering which regions governed essential functions like movement, speech, and memory. These intrepid pioneers laid the foundation for what would one day blossom into the field of brain-machine interfaces—a union of biology and technology poised to redefine human potential.

A pivotal breakthrough in this odyssey came with the ability to measure the brain's electrical symphony. The brain, an intricate web of neurons, communicates through rapid-fire electrical impulses, forming the foundation of thought, sensation, and action. By placing electrodes on the scalp, scientists began to "eavesdrop" on these neural conversations, gaining unprecedented insights into the brain's inner workings. This newfound ability to record brain activity unlocked doors not just to understanding the mind but also to engaging with it in profound and transformative ways.

Yet, merely listening to the brain's electrical whispers was not enough. The true challenge lay in establishing a two-way dialogue — bridging the divide between thought and action, mind and machine. This pursuit spurred the creation of ever-more sophisticated technologies capable of not only monitoring neural activity but also influencing it. The leap from passive observation to dynamic interaction marked the dawn of direct brain control over external devices, a step closer to realizing the dream of seamless human-machine integration.

One of the earliest triumphs of this integration was the cochlear implant, a device that has restored hearing to countless individuals. By directly stimulating the auditory nerve, cochlear implants exemplify the power of technology to reconnect the brain with sensory experiences once thought lost forever. But they are just the opening act in a much larger narrative. The same principles that allow these implants to deliver sound are now being harnessed to enable movement for the paralyzed, vision for the blind, and even thought-driven communication. Each development brings us closer to a future where the barriers between mind and machine dissolve, and the possibilities for human achievement expand beyond imagination.

～♪

The fusion of the brain with machines reaches far beyond the realm of medical rehabilitation, unveiling possibilities that stretch the limits of imagination. As these technologies advance, they promise not only to restore lost functions but to amplify human capabilities in extraordinary ways. We are on the cusp of an era where the melding of mind and machine offers a pathway to human augmentation, enabling feats that surpass the boundaries of our biological design.

Picture a world where thought alone commands intricate machinery—where robotic limbs, vehicles, or even entire systems respond instantaneously to the mind's directives. For those with physical disabilities, such advances could transform dependence into newfound autonomy, turning barriers into bridges. Yet, this potential is not confined to disability. Envision surgeons performing delicate operations with robotic precision, their thoughts guiding instruments more adeptly than hands ever could. Imagine athletes harnessing brain-machine interfaces to perfect their movements, pushing their performance to unprecedented heights in real time.

Though such visions may seem drawn from the pages of science fiction, they are firmly rooted in emerging scientific breakthroughs. The brain's remarkable plasticity—the ability to adapt, rewire, and learn—provides the foundation for this transformative potential. Just as the brain learns to command the muscles of the body, it can master the control of external devices. With time and practice, these interactions become second nature, creating a seamless connection between human intent and technological action.

The promise of brain-machine fusion extends beyond physical augmentation. Cognitive enhancement is a frontier of equal intrigue, offering profound implications for how we think, learn, and create. By connecting the brain to powerful computational systems, we could extend our intellectual reach, unlocking capabilities once deemed impossible. Memory could become boundless, attention sharper, and problem-solving more intuitive as machines take on part of the cognitive load. As the brain adapts to this partnership with artificial intelligence, entirely new ways of thinking might emerge, transforming not only what we can achieve but how we perceive the world and ourselves.

～

Artificial intelligence is the vital thread weaving together the intricate tapestry of brain-machine fusion. It serves as a translator, decoding the brain's intricate symphony of electrical signals into commands intelligible to machines. Without AI's interpretive power, the neural activity of the brain would remain a chaotic hum, incomprehensible to computers. With sophisticated algorithms, however, AI brings clarity to this neural noise, enabling direct communication between mind and machine—a feat once relegated to the realm of dreams.

The relationship between AI and the human brain is profoundly symbiotic, forming a partnership where both entities evolve together. AI learns to navigate the nuances of each individual's unique neural patterns, tailoring its responses with precision. Simultaneously, the brain adapts to this digital collaborator, honing its ability to control machines with increasing fluidity and confidence. This dynamic feedback loop lies at the core of brain-machine integration, fostering a seamless interplay that once seemed impossible.

As AI grows more advanced, its ability to understand and even predict human behavior reaches remarkable heights. This dual-edged sword offers both unprecedented opportunities and profound challenges. On one hand, it paves the way for brain-machine interfaces that are intuitive and responsive, finely tuned to the user's intentions. On the other, it raises pivotal questions about autonomy and control. If an AI anticipates your thoughts or actions, is it merely following your lead—or subtly guiding you? These questions probe deeply into the essence of free will and the very fabric of human decision-making.

For much of history, intelligence was seen as a uniquely biological trait, an innate hallmark of life. Machines, regardless of their complexity, were tools—extensions of human ingenuity but not entities capable of thought. The advent of AI has fundamentally reshaped this narrative. Machines now learn, adapt, and even create, erasing the once-clear boundary between human and artificial intelligence. No longer just tools, AI systems are emerging as collaborators—partners in solving problems, making decisions, and amplifying human cognitive abilities. As AI continues to evolve, it invites us to reconsider what it means to think, to innovate, and to exist in a world where intelligence is no longer confined to flesh and bone.

~*

This brings us to one of the most compelling and contentious ideas in the realm of brain-machine fusion: the singularity. Made famous by futurist Ray Kurzweil, the singularity envisions a moment when technological progress accelerates beyond human control, ushering in profound and unpredictable transformations to civilization. At its heart, the singularity represents the point where artificial intelligence surpasses human intelligence, entering an era where machines can enhance and replicate themselves without human oversight.

To Kurzweil and others who champion this concept, the singularity is not a distant possibility but an inevitable consequence of exponential technological growth. With each iteration, intelligent machines will refine and build more advanced versions of themselves, initiating a feedback loop of self-improvement. This cycle could spark advances at a pace far beyond human capability, propelling us into a future we can scarcely comprehend.

For Kurzweil, the singularity is not a dystopian scenario to dread but a transformative milestone to embrace. He imagines a world where humans and machines converge, forming a hybrid existence that transcends the limitations of biology. In this vision, the boundaries between mind and machine, body and technology, dissolve into seamless integration. Humans might upload their consciousness into digital frameworks, achieving a form of immortality, or enhance their minds with AI, unlocking cognitive abilities far beyond those we possess today.

The potential of such a future is staggering. Machines with superhuman intelligence could tackle humanity's most enduring challenges—curing diseases, eradicating poverty, and reversing environmental damage—at a scale and speed previously unimaginable. The singularity could herald a new golden age, where technology elevates every aspect of human life and reshapes our understanding of what it means to thrive.

Yet this vision is not without peril. The rise of intelligence beyond human comprehension introduces significant risks. Machines more intelligent than humans could act in ways that are unpredictable, or even harmful, to their creators. The prospect of losing control over these technologies is no longer a distant science-fiction trope but a pressing concern. As AI becomes increasingly autonomous, ensuring it aligns with human values and intentions becomes an intricate and critical challenge.

The singularity also forces us to confront profound ethical dilemmas. If we merge with machines, what becomes of our identity? Will we remain recognizably human, or will we evolve into something entirely new? These are not merely speculative questions but foundational issues that strike at the core of our existence. As the lines between mind and machine grow fainter, we must grapple with what it means to be human in an era defined by such transformative fusion.

∼

The fusion of the human brain with machines transcends mere technological innovation, delving deep into the realms of ethics and philosophy. The ability to interface directly with the mind forces us to confront profound questions about autonomy, privacy, and the very nature of consciousness. As we edge closer to a future where the boundaries between mind and machine blur, navigating these issues will be vital to charting an ethical path through this unprecedented frontier.

One of the most urgent ethical dilemmas is the question of autonomy. If machines can be guided by our thoughts, what becomes of free will? In a world where artificial intelligence can anticipate and even influence our decisions before we are consciously aware of them, are we still the authors of our own actions? This strikes at the core of what it means to be human. Throughout history, free will has been cherished as a defining trait of humanity, setting us apart from both animals and machines. Yet, as brain-machine interfaces evolve, this distinction grows increasingly ambiguous, challenging the very foundation of our identity.

Equally pressing is the question of responsibility. When actions are mediated by a brain-machine interface, who bears accountability? Is it the individual, the technology, or the engineers behind the system? This is not a theoretical concern but one with real-world implications for law, morality, and societal norms. As technology becomes an extension of the self, disentangling human agency from machine influence becomes a complex, and potentially contentious, endeavor.

Privacy, too, stands on a precarious precipice. For as long as humans have existed, our innermost thoughts have been our own—safe, private, and inviolable unless shared by choice. Brain-machine interfaces threaten to upend this sacred solitude. If machines can read and decode our thoughts, who will control that information? Could it be exploited, commercialized, or weaponized against us? The possibility of our mental landscapes being breached raises the specter of a future where even our most intimate selves are no longer free from intrusion.

Safeguarding mental privacy will be one of the defining ethical challenges of this era. New laws, technologies, and societal norms must emerge to protect the sanctity of the mind, ensuring our thoughts remain inviolate. At the heart of this challenge is a collective commitment to the idea that the mind is not merely a machine's data source but a sacred space deserving of respect and protection.

Perhaps the most enigmatic issue posed by this fusion is the question of consciousness itself. If a machine can replicate or enhance human thought, where does the machine end and the human begin? Could we, in time, transfer our consciousness into digital frameworks, achieving a form of immortality? What was once the stuff of speculative fiction is now a topic of serious scientific and philosophical inquiry.

Consciousness, the essence of our awareness and the core of our experience, is the most profound frontier of all. It defines our sense of self, our connection to the world, and our understanding of life and death. If machines could achieve consciousness, or if we could transplant our own into them, how would this reshape our perceptions of existence? These questions defy easy answers, yet they demand our attention as we advance toward an era where the fusion of mind and machine could redefine the essence of what it means to be alive.

The fusion of the human brain with machines will ripple through society, transforming the very fabric of our existence. As these technologies become more pervasive, they will redefine how we live, work, and connect with one another. While some of these shifts will bring unprecedented opportunities and advances, others may challenge the foundations of our social structures and values.

One of the most profound impacts will be felt in the workforce. As machines grow more intelligent and capable, they will increasingly take over tasks once performed by humans. This could result in greater efficiency and productivity, unlocking new potentials for innovation. Yet, it also brings the specter of job displacement and economic inequality. The rise of brain-machine interfaces could deepen these divides, with those who have access to such technologies gaining a significant edge in the job market, further widening the gap between the haves and the have-nots.

However, this fusion also carries the promise of new opportunities for work, creativity, and expression. By augmenting human abilities, brain-machine interfaces could empower us to achieve feats once thought impossible. Entirely new professions, artistic movements, and ways of thinking could emerge, reshaping the landscape of human endeavor. The challenge will lie in ensuring that these opportunities are accessible to all, and that the benefits of this fusion are distributed equitably across society.

As our cognitive abilities are enhanced, the education system may need to undergo a radical reimagining. Traditional models, which emphasize rote memorization and knowledge accumulation, may no longer suffice in a world where knowledge is effortlessly accessible through brain-machine interfaces. Instead, education could shift towards fostering skills needed to collaborate with intelligent machines, with critical thinking, creativity, and emotional intelligence taking center stage in shaping a new generation of learners.

The fusion of mind and machine will also profoundly alter our social relationships and sense of community. As we gain the ability to communicate directly mind-to-mind, the traditional forms of verbal and written communication could begin to fade into obsolescence. This might lead to a deeper, more intuitive understanding between people, where thoughts and experiences can be shared instantaneously. Yet, it could also create new forms of isolation and misunderstanding, as the boundaries between the public and private realms become increasingly porous.

Perhaps the most fundamental question will concern our identity itself. If we can upload our consciousness into machines, or enhance our minds with artificial intelligence, what does it mean to be human? Will we still see ourselves as part of a shared human experience, or will we begin to divide into new categories based on the extent of our technological augmentations? These profound questions will challenge our conceptions of belonging, community, and the very nature of what it means to be human in an increasingly integrated world of mind and machine.

~~

As we stand on the threshold of a new era, the fusion of the human brain with machines offers both remarkable opportunities and formidable challenges. The path ahead is shrouded in uncertainty, yet it is brimming with potential. The choices we make today will shape the future of humanity, determining how we live, work, and interact with one another in the coming years.

One of the most pressing challenges will be ensuring that the benefits of brain-machine fusion are accessible to all. As with any transformative technology, there is a risk that its advantages will be concentrated in the hands of a privileged few, exacerbating inequality and deepening social divides. To avoid this, we must craft policies and frameworks that guarantee equitable access to these advances, ensuring that everyone, regardless of circumstance, has the opportunity to partake in the benefits of this brave new world.

Equally important will be managing the inherent risks of this fusion. The potential for misuse or unforeseen consequences is significant, and it is imperative that we approach these technologies with caution and foresight. This will require the development of technical solutions, ethical guidelines, and robust legal frameworks that protect individual rights, ensure privacy, and foster the responsible use of these powerful tools.

At the same time, we must remain open to the boundless possibilities that this future holds. We are witnessing an age of unparalleled innovation, and the fusion of mind and machine has the power to transform our lives in ways we can scarcely imagine. By embracing this technology with optimism and a sense of responsibility, we have the opportunity to create a future that is not only groundbreaking but also humane—one where technology enhances the human experience, and its risks are managed with wisdom and care.

Fusing the human brain with machines is not a far-off dream; it is happening right now. We stand at the dawn of a journey that will fundamentally alter our understanding of who we are and how we relate to the world around us. The choices we make today will chart the course of this transformative journey, shaping the future of humanity for generations to come.

As we move forward, it is crucial to remember that this fusion is not just a technological challenge; it is, at its core, a human one. It calls for a deeper understanding of our identity, our values, and the future we wish to build. The decisions we make now will not only determine the trajectory of technology but also the destiny of humanity itself.

This book seeks to explore these profound ideas—an effort to unravel where we are, where we are headed, and what these developments mean for us as a species. We will delve into the science and technology that are making brain-machine fusion possible, confront the ethical and philosophical questions that arise from it, and consider the vast array of futures that might unfold.

As we embark on this journey, we must do so with a sense of both excitement and caution. The fusion of mind and machine is not merely the next step in technological evolution; it is a defining moment in human history. This convergence holds the power to address some of our greatest challenges, enhance human capabilities, and foster a brighter, more interconnected future for all. Yet, it also carries risks that must be managed with care to ensure that the benefits of this fusion are shared by all.

The future is unfolding more quickly than we might realize, and it will be shaped by our ability to navigate the intricate and often uncertain path of brain-machine fusion. Let us approach this future with both anticipation and prudence, recognizing that this moment represents not just another leap in technology but a profound shift in the course of human history.

CHAPTER 1

Journeys of Fusion

❦

The fusion of the human brain with machines is no longer a theoretical concept or a distant dream; it is a reality that is already transforming lives today. In this chapter, we will explore the personal journeys of three individuals—Gert-Jan Oskam, Matthew Angle, and Jens Naumann—whose experiences with brain-machine interfaces have profoundly impacted their lives. Their stories serve as powerful illustrations of the immense potential of this technology and how it is gradually blurring the lines between mind, body, and machine.

～

Gert-Jan Oskam's life was irrevocably changed in 2011 when a devastating bicycle accident left him paralyzed from the neck down. The accident severed his spinal cord, cutting off the communication between his brain and the muscles in his body. Like many others who suffer spinal cord injuries, Gert-Jan was told he would never walk again. In that instant, his independence, mobility, and sense of self were altered forever.

Yet, Gert-Jan's journey didn't end with his injury. It marked the beginning of an extraordinary chapter, one that would see him regain the ability to walk, thanks to the remarkable fusion of his mind with cutting-edge technology. In 2023, after years of living with paralysis, Gert-Jan became one of the first individuals to walk again using a brain-machine interface—a groundbreaking achievement made possible by a combination of advanced neuroscience, engineering, and sheer determination.

The technology that enabled Gert-Jan to walk again is the culmination of decades of research into how the brain controls movement. Essentially, the system creates a new communication pathway between the brain and the body, bypassing the damaged spinal cord. It begins with a device implanted in Gert-Jan's brain that detects electrical signals from his motor cortex, the area responsible for planning and executing movement. These signals are decoded by a computer and transmitted to an implant near his spinal cord, which then stimulates the nerves controlling his leg muscles.

Though complex, this process allows Gert-Jan to think about walking, and for those thoughts to be translated into movement. The brain-machine interface bridges the gap created by his spinal injury, giving him a level of control over his body that was once thought impossible. His story is a powerful testament to how technology can restore lost functions, essentially rebuilding the connection between mind and body.

However, the road to this achievement was anything but easy. Gert-Jan's journey was filled with years of physical therapy, setbacks, and uncertainty. The technology that would eventually allow him to walk was still in its experimental stages when he began. He underwent extensive training sessions to learn how to use the interface, as the brain, while highly adaptable, needs time to master new tools—especially one as complex as a brain-machine interface.

Gert-Jan's story isn't just about the technology—it's about the human spirit. It's a story of resilience, hope, and the willingness to push past what was once thought to be an insurmountable barrier. His experience reminds us that while brain-machine interfaces are remarkable tools, they require the mind's collaboration to unlock their full potential. For Gert-Jan, the fusion of mind and machine wasn't a replacement for human will—it was an extension of it.

Gert-Jan's remarkable journey shows us the transformative power of brain-machine fusion. It demonstrates that even when the body fails, the mind—empowered by technology—can find new ways to express itself. His story offers a hopeful glimpse into a future where the limitations of the body may no longer stand in the way of achieving independence and mobility.

～

While Gert-Jan's story is one of personal triumph, the story of Matthew Angle highlights the scientific and entrepreneurial journey behind such breakthroughs. Matthew Angle is the founder of Paradromics, a company at the forefront of developing high-bandwidth brain-machine interfaces. His work is not just about helping individuals regain lost functions—it's about pushing the boundaries of what these interfaces can achieve on a far broader scale.

Matthew's journey into the world of brain-machine interfaces began with a simple yet profound question: How can we build a better interface between the brain and technology? This inquiry led to the development of technologies that could enable seamless communication between the brain and machines. Unlike the interfaces that helped Gert-Jan walk again, which were primarily focused on restoring specific functions, Matthew's work aims to create versatile platforms that could be used across a wide range of applications—from medical rehabilitation to enhancing human cognition.

One of the key challenges in brain-machine interfaces is bandwidth: the capacity to transmit large amounts of data between the brain and machines. The brain is a remarkably complex organ, capable of processing vast amounts of information in parallel. However, current interfaces are limited in the amount of information they can capture and transmit. Matthew's work at Paradromics aims to overcome this limitation by developing interfaces capable of much higher data rates, enabling more precise and nuanced communication between mind and machine.

Paradromics' technology uses high-density electrode arrays that can be implanted in the brain, allowing them to record the activity of thousands of neurons simultaneously. This data-rich approach offers a much clearer picture of the brain's activity than previous technologies, enabling more accurate control of external devices and even the decoding of complex thoughts and intentions.

The potential applications for this technology are vast. In medicine, it could lead to new treatments for conditions like epilepsy, depression, and neurodegenerative diseases. For individuals with severe disabilities, it could open new pathways for communication and interaction with the world. Beyond medicine, it could lead to the development of new forms of human augmentation, enhancing cognitive abilities and even changing how we think.

Matthew's work represents a different kind of fusion—one that seeks to expand the boundaries of human cognition and creativity. His vision is one in which brain-machine interfaces become platforms for innovation, not just rehabilitation. By creating more powerful connections between the brain and technology, he is helping to unlock new possibilities for human enhancement.

However, Matthew's journey is not without challenges. Developing high-bandwidth brain-machine interfaces requires not only scientific innovation but careful ethical and societal considerations. The ability to decode and interpret brain activity raises profound questions about privacy, autonomy, and the nature of consciousness. Matthew has been an outspoken advocate for responsible development in the field, emphasizing the need for ethical frameworks to guide the advance of these technologies.

Matthew's story underscores the fact that the fusion of mind and machine is not only a technical challenge — it's also a deeply human one. It requires vision, creativity, and an understanding of the ethical dimensions of this work. As brain-machine interfaces become more widespread, the work of pioneers like Matthew will be critical to ensuring this technology benefits humanity.

~

While Gert-Jan and Matthew's stories focus on restoring and enhancing human abilities, Jens Naumann's journey highlights another crucial aspect of brain-machine fusion: sensory restoration. Jens was once blind, but thanks to a pioneering brain-machine interface, he became one of the first individuals in the world to receive a bionic eye.

Jens lost his sight in two accidents. The first, a mechanical mishap, left him partially blind; the second, a farm accident, took the remainder of his vision. For Jens, the loss of sight was devastating—not only did it rob him of his independence, but it also forever altered his perception of the world.

However, Jens refused to let blindness define his life. He sought out experimental treatments and eventually became a candidate for a brain-machine interface designed to restore vision. The technology he received works similarly to a cochlear implant but stimulates the visual cortex—the area of the brain responsible for processing visual information.

The system uses a camera mounted on glasses to capture images, which are then processed and converted into electrical signals that are sent to electrodes implanted in Jens' visual cortex. These signals stimulate specific neurons to create rudimentary vision, allowing Jens to perceive shapes, patterns, and even movement.

While the technology didn't restore perfect sight, it was a profound shift, enabling Jens to regain a level of independence and connection to the world that had been lost. His journey illustrates how brain-machine interfaces can not only restore functions but change the way we experience the world.

Jens' experience also highlights the challenges and limitations of current technology. While his vision is a remarkable achievement, it remains far from natural. The complexity of the brain makes replicating the full richness of sensory experience a daunting challenge. Jens' story reminds us that while brain-machine fusion holds immense potential, there is still much work to be done.

Despite these limitations, Jens' story is one of perseverance and hope. It shows the incredible power of technology to change lives, and it speaks to the adaptability of the human spirit in embracing new ways of experiencing the world.

~⁓

For Casey Harrell, the fusion of mind and machine became more than a technological marvel — it became his means of reconnecting with the world after ALS had taken away his ability to speak and move. ALS, a devastating neurodegenerative disease, gradually robs individuals of their muscle function, leaving them trapped inside their own bodies, fully aware but unable to communicate or control their surroundings. For Casey, who had spent his life advocating for environmental causes, the onset of ALS was particularly cruel, as it took away his voice in the most literal sense.

As his condition progressed, Casey found himself increasingly isolated, unable to engage with the world as he once had. Yet, even as his body failed him, his mind remained sharp, filled with the same passion and determination that had driven him throughout his life. It was this determination that led him to explore the possibilities of brain-machine interfaces — a technology that offered a glimmer of hope in an otherwise bleak situation.

Casey's journey with brain-machine interfaces began when he was introduced to a system that allowed him to communicate using only his thoughts. The technology, developed through years of research and innovation, involved implanting a device in Casey's brain that could interpret the electrical signals generated by his neurons. These signals, when decoded by a computer, were translated into text on a screen, giving Casey a way to express himself once again.

The process was not simple. It required extensive calibration, as the system needed to learn to recognize the specific patterns of neural activity associated with Casey's intended speech. This learning process was a two-way street—just as the machine had to learn to interpret Casey's brain signals, Casey had to learn to control the interface with his thoughts. Over time, however, the connection between mind and machine grew stronger, and Casey was able to communicate more effectively.

For Casey, the brain-machine interface was more than just a tool— it was a lifeline. It allowed him to reconnect with his loved ones, to share his thoughts and feelings, and to continue advocating for the causes he cared about. Through this technology, Casey found a way to transcend the physical limitations imposed by ALS, using the power of his mind to overcome the barriers that his body had placed before him.

But Casey's story is also a testament to the resilience of the human spirit. The technology that enabled him to communicate again was remarkable, but it was Casey's determination and will to keep fighting that made it truly powerful. His journey highlights the profound impact that brain-machine interfaces can have on individuals with severe disabilities, offering them not just a way to regain lost functions, but a means of reclaiming their place in the world.

Casey's experience with brain-machine interfaces underscores the transformative potential of this technology. It shows us that even in the face of seemingly insurmountable challenges, the fusion of mind and machine can open up new possibilities, allowing individuals to connect, communicate, and live more fully. As we continue to explore and develop these technologies, Casey's story serves as a reminder of the profound human impact that lies at the heart of this field.

The fusion of mind and machine is not just about enhancing abilities or restoring lost functions—it's about providing a bridge between the physical and the mental, allowing individuals like Casey to overcome the limitations of their bodies and reconnect with the world around them. His journey is a powerful example of how technology can be harnessed to serve the most fundamental of human needs: the need to communicate, to connect, and to live with dignity.

~

The stories of Gert-Jan Oskam, Matthew Angle, Jens Naumann, and Casey Harrell are each unique, yet they share a common thread: the fusion of the human mind with technology has the power to transform lives. Whether it's regaining the ability to walk, pushing the boundaries of what brain-machine interfaces can achieve, or restoring sight to the blind, these stories illustrate the incredible potential of this technology.

But these stories also remind us that brain-machine fusion is not just about technology—it's about people. It's about the individuals whose lives are changed by these interfaces, and the scientists and engineers who make these changes possible. It's about the resilience, determination, and creativity that drive this field forward. And it's about the ethical and societal questions that we must grapple with as we continue to explore the possibilities of this technology.

As we look to the future, the stories of Gert-Jan, Matthew, Jens, and Casey offer both inspiration and a reminder of the challenges ahead. They show us what is possible when we combine the power of the human mind with the potential of technology. But they also remind us that the journey is far from over. There are still many questions to be answered, many challenges to be overcome, and many lives to be transformed.

The fusion of mind and machine is not just a technical achievement—it is a human one. It is a testament to our ability to push beyond the limits of what was once thought possible, to create new ways of living and interacting with the world. As we continue to explore the possibilities of brain-machine interfaces, we must remember that at the heart of this journey are people—people like Gert-Jan, Matthew, Jens, and Casey—whose lives are forever changed by this remarkable technology.

But how did we arrive at this point? The personal stories of those who have experienced brain-machine interfaces firsthand are the latest chapters in a much longer narrative. To fully understand the present and future of this field, we must first look back at its origins. The next chapter will take us on a journey through the history of brain-machine interfaces, tracing the key milestones and discoveries that have brought us to the cusp of this new era of human-machine fusion.

CHAPTER 2

The Evolution of Fusion: The History of Brain-Machine Interfaces

The concept of fusing the human brain with machines may seem like a vision of our high-tech era, but its origins trace back much further. The journey from abstract ideas to today's advanced brain-machine interfaces is a tale woven with curiosity, innovation, and an unyielding drive to unlock the mysteries of the mind. To fully understand where we stand now—and where we are headed in the future—it is essential to look back at the pivotal milestones and discoveries that have shaped this field. The history of brain-machine interfaces is not merely a chronology of technological achievements; it is a reflection of humanity's enduring quest to bridge the divide between mind and machine, a journey that has evolved from the speculative musings of ancient philosophers to the groundbreaking innovations of the 21st century.

Long before the first electrode was implanted in a human brain, the connection between mind and body was a subject of intense curiosity and philosophical inquiry. Ancient civilizations were captivated by the mind's role in governing the body, often linking mental functions to the soul or other supernatural forces. The earliest known investigations into this relationship can be traced to ancient Egypt and Greece, where thinkers such as Imhotep and

Hippocrates set the stage for what would later evolve into the science of neuroscience.

In ancient Egypt, the brain was not initially regarded as the seat of consciousness or intellect. Instead, it was the heart that held dominion over thought and emotion. This belief was so deeply ingrained in Egyptian culture that, during mummification, the brain was often discarded, while the heart was carefully preserved. Yet, this perspective began to shift with the contributions of Greek philosophers like Alcmaeon of Croton, who began to question and challenge these long-held beliefs.

Alcmaeon, a pre-Socratic philosopher, was one of the first to suggest that the brain, not the heart, was the organ responsible for thought and sensation. His pioneering ideas were further developed by Hippocrates, who posited that the brain was not only the seat of intelligence but also the origin of all mental processes. These early philosophical insights laid the groundwork for a scientific understanding of the brain, offering the first inklings of how the mind and body might be interconnected.

While the notion of controlling the body through the mind—or even extending the mind's influence beyond the body—was not entirely foreign to these ancient thinkers, the technology to make such control a reality was still far beyond their reach. It would take centuries of scientific progress before humanity could begin to transform these philosophical speculations into tangible, practical advances.

～

The Enlightenment period ignited a renewed fascination with the study of the brain, driven by the era's emphasis on reason, science, and empirical observation. This intellectual revolution marked the dawn of modern neuroscience, as scholars began to dissect the brain's structure and function with a level of detail previously unimaginable.

One of the foremost thinkers of this era was René Descartes, the French philosopher and mathematician who famously declared, "Cogito, ergo sum" ("I think, therefore I am"). Descartes was deeply intrigued by the relationship between mind and body, and he proposed a dualistic model in which the mind and body were separate entities, interacting through the pineal gland. While his theory about the pineal gland was ultimately proven incorrect, Descartes' work was pivotal in framing the mind-body problem, setting the stage for future scientific exploration into how these two realms—mental and physical—might interact.

Throughout the 18th century, the study of the brain took on a more empirical, systematic approach. Scientists began to map the brain's anatomy with increasing precision, identifying distinct regions responsible for various functions. Among the most influential discoveries was that of Luigi Galvani, whose experiments in the late 1700s revealed that electrical stimulation could induce muscle contractions. This discovery led to the concept of "animal electricity" and the recognition that electrical signals played an essential role in the nervous system.

Galvani's groundbreaking work paved the way for others, including Alessandro Volta, who expanded on these findings to create the first battery. Volta's innovations in electricity and his exploration of its effects on the body laid the crucial groundwork for the use of electrical stimulation in both medicine and neuroscience. These early studies would ultimately contribute to the development of brain-machine interfaces, bringing us closer to the integration of mind and machine.

~❧

The 19th century was a period of remarkable progress in our understanding of the brain, driven by the efforts of scientists who endeavored to map its functions with greater precision. This era also saw the rise of phrenology, a now-discredited theory that suggested the shape of the skull could reveal a person's character and mental abilities. While phrenology has since been debunked, it contributed to the foundational idea that specific areas of the brain are responsible for distinct functions—an idea that would later be validated in a more sophisticated form.

Among the most significant advances of this century was the work of Pierre Paul Broca, a French physician who made groundbreaking discoveries about the localization of brain function. Broca identified that damage to a particular region of the brain (now known as Broca's area) led to the loss of speech production, a condition known as aphasia. This finding provided some of the first concrete evidence that the brain's different regions serve specialized roles, a concept that would become central to the field of neuroscience.

Around the same time, German scientists Gustav Fritsch and Eduard Hitzig conducted pioneering experiments on dogs' brains. By applying electrical stimulation to different areas of the brain, they were able to induce specific movements in corresponding parts of the body. Their work was instrumental in demonstrating the brain's role in controlling voluntary movement and in solidifying the principle of localization of function—the idea that different regions of the brain are responsible for distinct physiological tasks.

The 19th century also witnessed the birth of neurosurgery, a field that would play a crucial role in the development of brain-machine interfaces. In 1885, Scottish surgeon William Macewen performed the first successful brain surgery, removing a tumor from a patient's brain. This achievement marked a significant milestone, opening the door to more invasive techniques for studying and treating the brain, and laying the groundwork for future technologies designed to interact directly with neural tissue.

~⁀

The 20th century marked a period of extraordinary progress in brain research and the birth of cybernetics, a field that would become instrumental in the development of brain-machine interfaces. Coined by mathematician Norbert Wiener in 1948, cybernetics is the study of control and communication in both animals and machines. It provided a theoretical framework for understanding how systems—especially the brain—could be regulated and how information could flow between different components, whether biological or mechanical.

Wiener's pioneering work in cybernetics was profoundly influenced by the rise of computers and the growing understanding of how the brain processes information. He envisioned a future in which machines could interact with the brain in ways that mirrored the natural communication between neurons, laying the intellectual foundation for the development of brain-machine interfaces.

Simultaneously, breakthroughs in electrophysiology allowed scientists to record and manipulate the electrical activity of neurons with remarkable precision. A landmark achievement came in the 1920s when German neurologist Hans Berger invented the electroencephalogram (EEG). The EEG enabled researchers to non-invasively measure the brain's electrical activity, offering a new window into its inner workings. This technology would become an indispensable tool in brain-machine interface research, enabling real-time monitoring of brain activity and the development of techniques to interpret and harness these signals.

The mid-20th century also witnessed the first experiments in direct brain-machine interaction. In 1957, American neuroscientist José Delgado conducted a series of groundbreaking studies in which he implanted electrodes into the brains of animals, including monkeys and cats, to observe and influence their behavior. Delgado's most famous experiment involved a bull, where he used a remote-controlled device to halt the animal in its tracks by stimulating its brain. These experiments demonstrated that it was possible to control animal behavior through direct brain stimulation, opening the door to future attempts to apply similar techniques to human brains.

While controversial, Delgado's work was a precursor to the more refined brain-machine interfaces that would follow. His research showed that the brain could be influenced and even controlled through electrical stimulation—a principle that would become central to developing technologies aimed at restoring lost functions and enhancing human abilities.

The late 20th century marked a pivotal shift from theoretical and experimental research to the development of practical brain-machine interfaces that could be applied to humans. Advances in computing, electronics, and neurosurgery converged, leading to the creation of the first generation of devices capable of interacting directly with the human brain.

One of the earliest and most impactful applications of brain-machine interfaces was the cochlear implant, developed in the 1970s. Cochlear implants work by bypassing the damaged parts of the ear and directly stimulating the auditory nerve, allowing individuals with severe hearing loss to hear. The development of this technology was a landmark achievement, providing solid proof that it was possible to interface with the human nervous system in a way that could restore a lost sense, offering new hope to those who had once been unable to perceive sound.

Around the same time, researchers began to explore how brain-machine interfaces could restore movement to individuals with paralysis. A key figure in this field was neuroscientist and engineer Philip Kennedy, who developed the first neural implant designed to allow paralyzed patients to communicate purely through thought. In the late 1990s, Kennedy's team successfully implanted an electrode array into the brain of a patient with locked-in syndrome, enabling him to control a computer cursor using only his thoughts. This was one of the first demonstrations of a brain-machine interface restoring communication in a human, marking a significant milestone in the field and paving the way for future advances.

The late 20th century also saw breakthroughs in understanding neuroplasticity—the brain's remarkable ability to reorganize itself in response to experience. This discovery was essential to the development of brain-machine interfaces, as it suggested that the brain could adapt and learn to control new tools and devices, even those as foreign as robotic limbs or computer interfaces. This insight gave rise to the belief that, with time and training, the brain could learn to operate increasingly complex technologies.

By the end of the 20th century, brain-machine interfaces had moved beyond the realm of research and experimentation and were beginning to be used in clinical settings. These innovations offered renewed hope to individuals with disabilities and opened up new possibilities for human enhancement. The stage was set for the 21st century, a period that would see rapid expansion, refinement, and broader application of these transformative technologies.

~●

The dawn of the 21st century ushered in a rapid acceleration in the development and application of brain-machine interfaces. With advances in technology and a deeper understanding of the brain, these devices became more sophisticated and effective, gradually transitioning from laboratory experiments to real-world applications. This shift has fundamentally altered how we think about the relationship between mind and machine, bringing us closer to a future where the boundaries between the two are increasingly blurred.

One of the most groundbreaking developments during this period was the use of brain-machine interfaces to restore movement to individuals with paralysis. Researchers at prestigious institutions such as Brown University and the University of Pittsburgh pioneered systems that enabled paralyzed patients to control robotic arms using only their thoughts. This innovation allowed them to perform basic tasks—like drinking from a cup or shaking hands—once thought impossible for individuals with paralysis. These advances were made possible by the creation of high-density electrode arrays and the development of more powerful algorithms to decode neural signals with greater precision, enabling smoother and more responsive control.

Simultaneously, the field of brain-machine interfaces expanded into new territories such as cognitive enhancement and communication. Systems like BrainGate, which allows individuals to control computers and other devices solely through their thoughts, offered new opportunities for those with severe disabilities. These interfaces not only improved the quality of life for users but also laid the foundation for future applications that could enhance cognitive abilities or facilitate direct mind-to-mind communication, opening up profound new possibilities for interaction.

Alongside these technological advances, the 21st century has seen a growing focus on the ethical and societal implications of brain-machine interfaces. As these technologies become more embedded in our daily lives, they raise pressing questions about privacy, autonomy, and the very nature of consciousness. The fusion of mind and machine is not merely a technical challenge—it represents a profound transformation in what it means to be human. As we continue to refine these technologies, it is crucial that we consider their impact not only on individuals but on society as a whole, ensuring that the integration of mind and machine serves the greater good and respects fundamental ethical principles.

～୨

The history of brain-machine interfaces is a journey from abstract concepts to practical realities, a testament to humanity's relentless drive to bridge the gap between mind and machine. It is a story of intellectual curiosity, scientific inquiry, and a deep desire to understand the brain's complexities and harness its power. From the philosophical musings of ancient thinkers to the groundbreaking experiments of the 20th century, each step has brought us closer to the ultimate goal of merging the mind with technology.

As we reflect on this journey, we can appreciate the remarkable progress made, but we also recognize the challenges that remain. The fusion of mind and machine is still in its early stages, and there is much left to discover. However, the journey is far from over. The strides of the past have laid a solid foundation for the future, and the next chapter in the history of brain-machine interfaces is still unfolding.

In the chapters that follow, we will delve into the current state of brain-machine interfaces, exploring the cutting-edge research propelling the field forward. We will examine the latest technologies, the hurdles they face, and the immense potential they hold to transform human life. The fusion of mind and machine is not a distant dream—it is a reality unfolding before us, and its implications will profoundly shape the trajectory of human history for generations to come.

CHAPTER 3

The Fusion Points: Reading the Brain's Signals with Electrodes

〜

Imagine sitting at a table, reaching out for a glass of water, but despite your focused effort, your hand remains still. For individuals living with paralysis or neurological disorders, even the simplest actions like this can seem impossible. This is where brain-machine interfaces come into play—systems that enable the brain to communicate directly with external devices, bypassing damaged nerves and translating thought into action. But to achieve this, we must first decode the brain's electrical signals. The key tools enabling this are electrodes, which serve as the vital bridge between human thought and machine response.

Electrodes work by capturing the brain's electrical activity, detecting signals from neurons—tiny cells that communicate with one another through electrical impulses. Each neuron sends out signals, and when enough of them fire in concert, they produce the electrical patterns we recognize as brainwaves. Electrodes convert these electrical impulses into data that machines can understand, enabling real-time control over robotic arms, computers, or even something as simple as a cursor on a screen.

However, capturing these signals isn't as straightforward as simply placing a device on the scalp. The quality of the signals depends on a variety of factors: the location of the electrodes, the number of electrodes used, and whether they are placed on the scalp, beneath it, or directly implanted into the brain. The deeper the electrodes are positioned, the clearer and more accurate the signals tend to be. But, deeper placements come with increased complexity, both in terms of the procedure and the potential risks.

~❧

Before delving into the technology of brain-machine interfaces, it's crucial to understand how the brain communicates. Neurons, the brain's messengers, transmit information to each other using electrical impulses, known as action potentials. When a neuron is activated, it sends an electrical signal down its axon to the next neuron, triggering a chain reaction of communication. Neurons don't work in isolation; they collaborate in vast networks, and the collective activity of these neurons generates brainwaves.

Brainwaves are categorized by their frequency—how quickly they oscillate—and each type of brainwave is linked to different mental states. For example:

- Delta waves are slow and occur during deep sleep.
- Theta waves are associated with drowsiness and light sleep.
- Alpha waves are common when you're awake but in a relaxed state.
- Beta waves are more prevalent when you're alert and focused.
- Gamma waves are fast and correspond to higher cognitive functions like problem-solving and complex thought.

Electrodes detect these brainwaves by capturing the electrical activity produced by large groups of neurons. The clearer and stronger the signal from these neurons, the easier it is for a machine to interpret what the brain is trying to convey. The challenge, however, lies in capturing these signals with enough precision and detail to ensure effective communication, all while minimizing invasiveness in the process.

The simplest and most accessible method for capturing brain activity is through non-invasive electrodes placed on the scalp. These electrodes are used in electroencephalography (EEG), a technique that has been in use for decades. EEG electrodes detect brainwaves through the scalp and are commonly used in medical diagnostics for conditions such as epilepsy or sleep disorders. However, EEG technology is also being adapted for brain-machine interfaces, enabling users to control devices simply by concentrating or relaxing.

One promising application of EEG technology is neurofeedback, particularly for children with Attention Deficit Disorder (ADD). In this system, children wear an EEG headset that measures their brain activity while engaging with a video game designed to improve concentration. The game responds to the child's brainwaves by rewarding them when their focus improves, creating a feedback loop that allows them to learn how to regulate their attention. This non-invasive, drug-free approach to managing ADD symptoms demonstrates how even relatively simple brainwave-reading technology can have a significant impact in the real world.

While EEG has the advantage of being easy to use and non-invasive, it does come with limitations. The skull and scalp tissues weaken the signals, making it difficult to capture the fine details of individual neuron activity. EEG is excellent for detecting broad brain states—such as whether someone is alert, relaxed, or asleep—but it struggles with tasks that require more precise motor control, such as moving a robotic arm or typing. To gather more detailed information from the brain, we need to go deeper.

~᭡

One step deeper than surface-level EEG is the use of subdermal electrodes, which are placed beneath the scalp but above the skull. These electrodes provide a middle ground between non-invasive techniques like EEG and fully invasive methods. Positioned closer to the brain, subdermal electrodes capture stronger signals than EEG without the need to penetrate the skull or brain tissue.

Subdermal electrodes are less invasive than fully implanted electrodes and don't require major surgery, but they do involve a minor procedure to place them beneath the scalp. This placement allows for more accurate readings of brain activity, offering clearer data for controlling devices or interpreting brain states.

While subdermal electrodes are not yet as widely used as EEG or more invasive methods, they are gaining attention for their potential in long-term monitoring and control. Because they are placed just under the skin, they can be worn comfortably for extended periods, making them suitable for applications such as seizure monitoring or even basic brain-machine interfaces. These interfaces may not require the extreme precision of deep brain implants but still benefit from the enhanced signal clarity provided by subdermal electrodes.

~᭡

When high precision is required, invasive electrodes are used. These electrodes are implanted directly into the brain tissue, allowing them to capture the electrical activity of individual neurons or small groups of neurons. This level of detail is essential for individuals who need fine motor control—such as those who are paralyzed and wish to control a robotic arm or a computer cursor.

One of the most commonly used invasive systems involves intracortical microelectrodes, which are inserted into the brain's cortex. The cortex is the region responsible for higher functions like movement, thought, and language. By being embedded directly into brain tissue, these electrodes can detect the electrical activity of individual neurons, enabling highly precise control over external devices.

BrainGate, a pioneering system in this field, uses these microelectrodes to assist individuals with severe motor impairments. In some cases, people who are unable to move their limbs can control computers or robotic limbs simply by thinking about moving. The electrodes capture the brain's intended commands, and a computer decodes these signals, translating them into instructions for the device.

However, the advantages of invasive electrodes come with significant challenges. The body often reacts to these implants by forming scar tissue around the electrodes, which can weaken the signal over time. Additionally, implanting these electrodes requires brain surgery, which carries inherent risks such as infection or brain injury. Because of these risks, invasive electrodes are typically reserved for individuals with severe disabilities, where the potential benefits outweigh the possible complications.

~

For those who need more precision than non-invasive methods but want to avoid the risks associated with deep brain surgery, electrocorticography (ECoG) provides a valuable middle ground. ECoG electrodes are placed on the surface of the brain, just beneath the skull but above the brain tissue itself. Positioned directly on the brain's surface, these electrodes capture clearer and more detailed signals than scalp-based EEG, while avoiding the need to penetrate brain tissue, thus reducing the risks tied to fully invasive implants.

Historically, ECoG electrodes have been used in clinical settings to monitor brain activity during surgeries, particularly for patients with epilepsy. However, researchers are now exploring their potential in brain-machine interfaces. Because ECoG provides higher-quality signals than EEG, it could be used to help individuals control devices like prosthetic limbs or communication tools. This offers a level of precision and control that is more refined than scalp-based methods, but without the invasive nature of deep brain implants.

~

One of the key factors in the effectiveness of any brain-machine system is electrode density—the number of electrodes placed in a given area of the brain. The more electrodes that can be packed into a small space, the more detailed the signal they can capture, leading to greater control over external devices.

Take Neuralink, for example, a company that has developed a system capable of implanting over 1,000 electrodes in a small region of the brain. These electrodes are designed to capture the activity of hundreds or even thousands of neurons, enabling highly detailed brain-machine communication. By using tiny, flexible threads that are thinner than a human hair, Neuralink can implant dense arrays of electrodes with minimal damage to brain tissue.

Why is this important? The brain is home to billions of neurons, and capturing signals from as many neurons as possible is crucial for translating thought into precise machine actions. With a higher density of electrodes, fewer neurons need to fire to generate a usable signal, making the system more responsive and accurate.

However, even with thousands of electrodes, challenges remain. The brain's electrical activity is extraordinarily complex, with many neurons firing simultaneously in different patterns. To achieve seamless brain-machine communication, both the density and precision of the electrodes must continue to improve. Researchers are exploring new frontiers in material science and nanotechnology to develop smaller, more flexible electrodes that can capture these signals without causing harm to the delicate brain tissue.

～

Looking ahead, the future of electrodes will focus on making them smaller, smarter, and more seamlessly integrated with the body. One of the most promising areas of development is wireless electrodes, which would eliminate the need for cumbersome wires connecting the brain to external devices. Wireless technology would offer users greater freedom of movement, making brain-machine systems more practical for everyday use and enabling more natural, fluid interactions with technology.

Another critical area of research is biocompatible materials. Traditional electrodes, often made from metals like platinum or gold, can cause irritation or inflammation in brain tissue, especially with prolonged use. To address this, new materials such as flexible polymers and graphene are being explored. These materials are more compatible with brain tissue, offering a better fit for the brain's dynamic, ever-moving environment. Flexible polymers and graphene can move with the brain, reducing the risk of tissue damage and maintaining long-term signal quality. This shift towards biocompatibility not only improves the safety and comfort of brain-machine interfaces but also enhances their potential for long-term, reliable use in a variety of applications.

～

Electrodes play a crucial role in capturing the brain's signals, but transforming those signals into meaningful actions requires another layer of technology: artificial intelligence (AI). AI is responsible for decoding the brain's intricate electrical patterns and converting them into commands that machines can understand and respond to. By analyzing the data captured by electrodes, AI algorithms can interpret the brain's intentions and translate them into precise actions, whether it's controlling a robotic arm, moving a cursor, or even facilitating communication.

In the next chapter, we will delve deeper into how AI works in tandem with electrodes to interpret brain signals. This collaboration is key to making brain-machine systems not only more efficient but also more intuitive and powerful, unlocking new possibilities for both medical and everyday applications.

CHAPTER 4

The Role of Artificial Intelligence in Merging Mind and Machine

I magine sitting in front of a computer, trying to move the cursor across the screen using only your thoughts. Electrodes are placed on your scalp, capturing the electrical signals emanating from your brain. These signals are a chaotic mix of brain activity, reflecting a multitude of thoughts and processes. So, how do we turn this complex data into the simple action of moving a cursor? This is where artificial intelligence (AI) becomes essential. AI is the key to interpreting the brain's electrical data and translating it into commands that machines can understand.

Without AI, the electrical impulses captured by the electrodes would simply be noise—patterns of activity that hold no meaning on their own. But with the power of AI, these signals are deciphered, transforming thoughts into actions and enabling seamless interaction between the brain and machines. In this chapter, we'll explore how AI collaborates with electrodes to enhance brain-machine interfaces, making them smarter and more intuitive, and why AI is critical to merging human intelligence with machines.

To understand AI's role in brain-machine interfaces, it's important to first grasp the problem it solves. The brain contains around 86 billion neurons, all communicating through electrical signals.

When we think, move, or even rest, groups of neurons fire off electrical impulses that travel throughout the brain and body. These impulses are what electrodes capture. However, while electrodes can detect the signals, they cannot inherently interpret what those signals mean.

Think of these signals as a foreign language. The brain's electrical activity is like a conversation in a dialect that no machine speaks. Neurons fire rapidly, with different areas of the brain using distinct "languages" depending on the task at hand. For example, the motor cortex, responsible for movement, sends a different type of signal than the visual cortex, which processes sight. Each region has its own firing patterns, and even within those areas, the signals vary depending on what thought or action is being processed.

AI's job is to act as a translator. It takes the raw electrical signals picked up by the electrodes and decodes them, identifying patterns that correspond to specific thoughts or actions. This decoding process enables brain-machine interfaces to function, allowing individuals to control prosthetic limbs, operate computers, or communicate through systems designed for those unable to speak. By interpreting these complex signals, AI makes it possible for machines to understand and respond to the brain's electrical activity, creating a bridge between thought and action.

～

The way AI learns to interpret brain signals is through a process called machine learning. Machine learning enables AI systems to learn from data, recognize patterns, and make predictions based on the information it has observed. In the context of brain-machine interfaces, the AI is trained on brain activity data while the user performs specific tasks. Over time, the AI becomes adept at identifying the neural patterns associated with those tasks.

Here's how this works in practice: A person might be asked to think about moving their hand to the left or right while their brain activity is recorded. The AI system analyzes the electrical patterns generated by the brain when the person thinks about making those movements. By repeating this process, the AI starts to recognize the distinct neural patterns that correspond to left or right hand movement. As this training continues, the AI improves its ability to predict the user's intended action based on their brain activity.

Once the AI is sufficiently trained, it can make real-time predictions about the user's intentions. This is the foundation of controlling machines through thought—AI translates the brain's electrical activity into actionable commands that the machine can follow, enabling seamless brain-machine interaction.

～

The process of training AI to interpret brain signals doesn't stop once the system has learned the basic patterns. The brain is constantly adapting, and for the AI to remain effective, it must adapt as well. This is where reinforcement learning comes into play. In reinforcement learning, the AI refines its understanding of brain signals by receiving feedback on its performance, allowing it to improve over time.

For example, if the AI is helping someone control a robotic arm, it might initially make errors, like misinterpreting the intention to move left as a command to move right. Each time the user corrects the system, the AI adjusts its interpretation. With continued interaction, the AI gets better at predicting the user's intentions and responding accurately. This ongoing feedback loop is essential for making AI a powerful tool in brain-machine interfaces, as it allows the system to become more responsive and accurate based on real-world use.

AI systems can be trained on an individual's unique brain data, learning to recognize their specific neural patterns and provide precise control over external devices. Since no two brains are identical, the neural patterns associated with specific actions can vary from person to person. AI's ability to adapt to these individual differences is what makes brain-machine interfaces so effective for a diverse range of users, ensuring that each system is tailored to the unique needs and signals of the individual.

~

A powerful form of AI known as deep learning has significantly advanced brain-machine interfaces. Deep learning utilizes neural networks, which are designed to mimic the structure of the human brain, to analyze complex data. This allows AI systems to process brain signals at a much deeper level, detecting patterns and nuances that simpler algorithms might overlook.

Deep learning has shown particular promise in decoding more complex brain functions, such as speech. In one groundbreaking project, researchers developed a deep learning system capable of translating brain signals associated with speech into actual words. This technology was tested on individuals who had lost the ability to speak due to neurological conditions. By using deep learning to decode the brain's speech-related signals, the system generated either text or synthesized speech, enabling these individuals to communicate simply by thinking about what they wanted to say.

The implications of this are immense. Individuals who have lost the ability to speak due to strokes, ALS, or other conditions could regain their voices through AI. As these systems process more data over time, they become faster and more accurate, further improving their ability to help those who are unable to communicate in traditional ways. The ability to "speak" through thought represents a profound leap forward in both medical technology and human communication.

~

AI has also made significant strides in decoding movement-related brain signals, which has immense potential for people with paralysis or other mobility impairments. The ability to control a prosthetic limb or robotic device with thought can be life-changing, but controlling something as intricate as an arm requires detailed information about what the brain is signaling—such as where to move, how fast, and how much force to apply.

This is where neural decoding models come into play. These AI systems predict movement based on the electrical signals emanating from the brain's motor cortex, the region responsible for controlling voluntary movement. Neural decoding models analyze brain activity and map out the patterns that correspond to specific movements, whether it's reaching for an object, flexing a muscle, or walking.

One compelling example comes from researchers who developed an AI system that decodes brain signals associated with walking. By analyzing these signals, the AI was able to control a robotic exoskeleton, enabling individuals who had lost the ability to walk to take steps again. The system predicted the user's intended movements in real-time, allowing them to regain mobility through thought.

This goes beyond merely restoring movement—it enhances mobility by giving users control over assistive devices. For example, someone using a robotic prosthetic limb could not only control basic movements but also adjust the pressure of their grip or the speed of their actions, all through brain signals. This level of control offers a new dimension of independence and empowerment for individuals with mobility challenges.

～～

AI systems have the remarkable ability to adapt to changes in brain activity, whether those changes are due to fatigue, stress, or the process of learning. This adaptability is crucial because brain activity is not static—it fluctuates based on a variety of factors, including emotional states, physical conditions, and cognitive load.

For example, someone using an AI-powered prosthetic limb may experience shifts in neural patterns as they become more comfortable with the device. As the user grows more familiar with the prosthetic, the AI system continuously learns from these changes, updating its understanding of the brain's signals to ensure the prosthetic remains responsive. This dynamic learning process is key to making brain-machine interfaces feel natural, as the system adjusts to the user's evolving needs and conditions.

AI systems can also quickly learn from new tasks. If a user wishes to switch from controlling a computer cursor to operating a drone, the AI can adapt to the new task by learning how to decode the specific brain signals associated with that action. This flexibility is what makes brain-machine interfaces so versatile, enabling them to be applied across a wide range of activities—from controlling devices in the home to performing more complex, high-level tasks. The ability of AI to adjust and learn continuously is what ensures that these systems remain practical and effective for a variety of real-world applications.

~

While AI is essential for interpreting brain signals and translating them into actions, it also holds significant potential for enhancing cognitive functions. Researchers are exploring how AI can not only read brain activity but also improve it, helping individuals think more clearly, remember better, or concentrate more deeply.

For instance, some researchers are experimenting with AI systems that stimulate specific areas of the brain to boost memory recall. By delivering targeted electrical stimulation to regions associated with memory, these systems help individuals retrieve information more effectively. In the future, AI could become a tool for cognitive enhancement, much like we use external devices to augment our physical abilities. These advances could offer solutions for individuals experiencing memory loss or cognitive decline, providing new opportunities for improved mental function.

Another area of exploration is using AI to improve attention and focus. AI could monitor brain activity in real-time and provide feedback to help users stay on task. This approach could be particularly beneficial for individuals with attention disorders, such as ADHD. By tracking when attention begins to wander, the system could deliver gentle stimulation or feedback, guiding the user back to their task. Such interventions could help people maintain focus in work or study environments, offering a potential non-invasive treatment for conditions that affect attention and concentration.

In these ways, AI is poised to go beyond just decoding brain signals—it could also play an active role in enhancing mental capabilities, allowing individuals to unlock their cognitive potential and manage brain-related challenges more effectively.

～

One of the most intriguing possibilities for AI in brain-machine interfaces is enabling brain-to-brain communication. Imagine being able to send thoughts or information directly from one person's brain to another without the need to speak or write. In this scenario, AI could serve as the intermediary, interpreting the neural signals from one brain and transmitting that information to another in real time.

In one study, researchers used brain-machine interfaces to allow two people to play a game together using only their brain signals. One person's brain activity was transmitted to the other person's brain through AI, enabling them to collaborate without any verbal or physical communication. This experiment opens up the potential for future technologies where humans could directly share thoughts, ideas, or even emotions through brain-to-brain communication, creating a new form of connection beyond traditional means.

If AI continues to advance in its ability to decode and interpret brain signals with greater precision, we may one day be able to bypass conventional forms of communication altogether. This could revolutionize how people interact, opening new avenues for empathy, collaboration, and understanding. The possibilities for human connection could expand beyond language barriers, offering a radically new way for people to share their inner experiences with one another.

~~~

The future of brain-machine interfaces is inextricably linked to the progress of AI. As AI systems become more advanced, they will enable increasingly seamless interactions between the brain and machines. We can envision a world where controlling devices with our minds feels as natural as moving our own limbs, breaking down the barriers between human intention and technological action.

AI also holds the potential to enhance cognitive functions, such as improving memory and attention, and even revolutionizing how we communicate. The merging of mind and machine will not be confined to physical movement—it will expand into the realms of thought, perception, and communication, allowing for deeper connections and more intuitive interactions with technology.

AI is the key to unlocking this future. As we continue to refine how AI decodes brain activity, we move closer to a reality where the boundaries between human thought and machine action blur, creating new opportunities for how we live, work, and interact.

In the next chapter, we will explore real-world applications of brain-machine interfaces, from helping people regain mobility to offering new forms of communication. These examples highlight how the fusion of AI and brain-machine interfaces is already transforming lives, while providing a glimpse into the incredible possibilities that lie ahead.

CHAPTER 5

Real-World Applications of Brain-Machine Interfaces

❧

The world of patient rehabilitation and prosthetics has undergone a remarkable transformation in recent years, thanks to the integration of brain-machine interfaces. These cutting-edge technologies are revolutionizing how we approach motor recovery and artificial limb control, rewriting the stories of countless individuals whose lives have been altered by injury or amputation.

∿

In the quiet of his Florida home, Johnny Matheny sat at the piano, his fingers hovering over the keys. The melody of "Amazing Grace" filled the air, each note a testament to human perseverance and technological marvel. For Johnny, this moment was more than just music - it was a symphony of hope, a crescendo of possibility that had been building since that fateful day in 2007 when cancer stole his left arm.

The journey to this moment had been long and arduous. For years after his amputation, Johnny grappled with the limitations of conventional prosthetics. They were clumsy, uncomfortable, and a constant reminder of what he had lost. But he was not one to accept defeat. His spirit, forged in the crucible of adversity, drove him to seek something more.

In 2018, that search led him to the cutting edge of prosthetic technology. Johnny became the first person to take home the Modular Prosthetic Limb (MPL), a marvel of engineering developed by the brilliant minds at Johns Hopkins Applied Physics Laboratory. This wasn't just a replacement for his lost arm - it was a gateway to a new world of possibilities.

As Johnny's fingers danced across the piano keys, each movement was a testament to the intricate symphony of technology at work within the MPL. Brushless torque motors whirred silently, translating his thoughts into action with a fluidity that seemed almost magical. Slotless brushless DC motors allowed for precise, delicate movements that could easily cradle an egg or grasp a pen.

But the true magic lay in the connection between man and machine. The MPL wasn't just responding to Johnny's physical movements but reading his very intentions. Through the miracle of myoelectric technology, the electrical signals generated by Johnny's remaining muscles were captured, interpreted, and translated into action. It was as if the arm could read his mind, responding to his thoughts with an intuitiveness that blurred the line between flesh and circuitry.

For Johnny, each day with the MPL was a new adventure, a chance to rediscover abilities he thought were lost forever. He could feel the soft fur of his dog, adjust the delicate tuning pegs on his guitar, and even type on a keyboard with both hands. These seemingly simple acts were profound victories, each a step towards reclaiming the life he once knew.

Yet, Johnny's journey was about more than personal triumph. He saw himself as a pioneer, blazing a trail for countless others who had lost limbs to injury or disease. Every challenge he faced, every hurdle he overcame, was a stepping stone towards a future where the loss of a limb no longer meant the loss of ability or independence.

As the last notes of "Amazing Grace" faded, Johnny sat back, a smile playing on his lips. The arm that once seemed like science fiction was now a part of him, a bridge between human intention and technological innovation. At that moment, as he flexed his prosthetic fingers and felt the responsive hum of motors and circuits, Johnny knew that he was living proof of an extraordinary truth: the future of prosthetics wasn't just coming - it was already here, and it was beautiful.

The journey to create these advanced prostheses is fraught with technical challenges. Engineers must achieve high torque and speed within a compact design while ensuring the device remains lightweight and easy to use. It's a delicate balance, like a biological limb's intricate dance of muscles and nerves.

A symphony of compact, high-performance motion control solutions lies at the core of these modern prostheses. Brushless torque motors, gear motors, and slotless brushless DC motors work in harmony, each playing a crucial role in mimicking the complex movements of a biological limb. These motors are engineering marvels in their own right, offering an impressive torque-to-weight ratio that allows for human-like strength in a lightweight package. Their precise control capabilities enable the fine motor movements necessary for delicate tasks. At the same time, their low noise and vibration profiles contribute to a more natural feel.

But motors alone do not make a limb. The control system of these prostheses is a sophisticated network of biosensors, mechanical sensors, and microprocessor-powered controllers. Biosensors act as the prosthesis's nervous system, detecting and interpreting the user's intention to move. Mechanical sensors provide constant feedback on the limb's position and movement, much like proprioception in a biological limb. At the heart of it all, a microprocessor serves as the brain, continuously monitoring and adjusting the prosthesis's movements in real time.

The Modular Prosthetic Limb (MPL) used by Johnny Matheny represents the pinnacle of this technology. Its anthropomorphic design not only mimics the appearance of a human arm but also replicates its strength and dexterity. High-resolution tactile and position-sensing capabilities allow for unprecedented control and feedback. The MPL's neural interface enables intuitive, closed-loop control, bringing us closer to the goal of true "prosthesis embodiment"—where the artificial limb feels like a natural extension of the user's body.

Delving deeper into the MPL's technology, we find the Local Motor Controller (LMC), a marvel of miniaturization. This compact component manages brushless DC motor commutation, sensor signal sampling, and communication with the central limb controller. It also monitors crucial parameters like joint temperature, torque, position, current, and rotor position - essential for smooth, natural movement.

The control of these advanced prostheses often relies on electromyography (EMG) signals - the electrical activity generated by muscle contractions. These signals are detected, interpreted, and translated into prosthetic movements. Some cutting-edge research combines EMG with real-time ultrasound imaging of muscle contractions, promising even more precise control in the future.

The control becomes even more intuitive for users like Matheny, who underwent Targeted Muscle Reinnervation (TMR) surgery. This groundbreaking procedure reroutes nerves that once controlled the amputated limb to other muscles, allowing users to control their prosthesis simply by thinking about moving their hand, just as they would with a biological limb.

Yet, control is only half the equation. True embodiment requires sensory feedback - the ability to feel what the prosthetic limb is touching. Advanced prostheses now incorporate sensors that can detect touch, pressure, and even temperature, sending this information back to the user's nervous system. This bidirectional communication between the user and the prosthesis represents a significant leap toward creating artificial limbs that truly feel like a body part.

These technological advances have opened up new possibilities for prosthesis users. Tasks that were once thought impossible, like playing a musical instrument, have become achievable. As Johnny Matheny's fingers danced across the piano keys, playing "Amazing Grace," he wasn't just making music - he was demonstrating the incredible potential of modern prosthetics.

The field of biomechatronics continues to evolve, driven by the collaborative efforts of engineers, neuroscientists, and medical professionals. With each breakthrough, we move closer to a future where losing a limb no longer means losing ability. The story of Johnny Matheny and the MPL is not just a tale of personal triumph - it's a glimpse into a future where the boundaries between biology and technology continue to blur, offering hope and possibility to millions worldwide.

~•

Interestingly, a study led by Dr. Linda Resnik of the Providence VA Medical Center found that user satisfaction only sometimes correlates with the complexity of the prosthetic device. The survey of over 400 Veteran users of upper-limb prostheses found little difference in satisfaction among users of myoelectric, body-powered, and cosmetic devices.

This highlights the importance of personalization and user-specific needs in prosthetic development. As these stories illustrate, these advances are not just about restoring function but about rekindling hope and redefining what's possible. With each technological leap, we're building better prostheses and rebuilding lives, one neural signal at a time. The future of rehabilitation is here, and it's more incredible than we ever imagined.

While motor rehabilitation has seen significant progress, sensory brain-computer interfaces are equally transformative. Cochlear implants, perhaps the most well-known sensory brain-computer interface, have restored hearing to individuals with severe to profound hearing loss for decades. These devices bypass damaged parts of the ear and directly stimulate the auditory nerve, allowing users to perceive sound.

In a quiet suburb of Boston, Emma Sullivan sat at her kitchen table, her fingers brushing the smooth surface of a device resembling a futuristic hearing aid. At 32, Emma had spent her life in silence, her cochlear nerves damaged beyond repair by a childhood illness. Today, in this imagined scenario, everything was about to change.

The audiologist carefully positioned the external component of a cochlear implant behind Emma's ear. Tension filled the room as Emma's husband, Mark, squeezed her hand, his eyes reflecting a mix of hope and apprehension. With a subtle nod, Emma signaled the audiologist to proceed.

As the device activated, the world seemed to pause. Emma's eyes widened, her lips parted in a silent gasp, and tears began to stream down her face. For the first time in decades, she could hear.

While Emma's story is hypothetical, the technologies driving moments like these are very real. Sensory brain-computer interfaces like cochlear implants have already transformed the lives of many by bypassing damaged sensory pathways. These devices convert sound waves into electrical signals, transmitting them directly to the auditory nerve, allowing individuals to perceive sound in new and profound ways.

The success of cochlear implants has inspired researchers worldwide to push the boundaries of what sensory brain-computer interfaces can achieve, including efforts to restore vision to the blind. A team led by Dr. Maria Asplund at Chalmers University of Technology in Sweden is developing a revolutionary new vision implant. Unlike earlier bulky devices, this implant uses electrodes the size of single neurons, arranged in a thread-like structure. These microscopic electrodes, designed to remain intact and functional in the body over time, represent a leap forward in vision restoration technology.

The implant bypasses the damaged parts of the eye and directly stimulates the visual cortex in the brain. Thousands of electrodes in the device can act as individual pixels, creating a grid of light patterns that form basic images. Though the resulting vision is not like natural sight, it offers the potential to perceive shapes, navigate spaces, and regain a sense of independence.

The innovation lies not just in the implant's miniature size but in its resilience. Built with a unique combination of materials, including a corrosion-resistant conducting polymer, the implant is designed to withstand the body's challenging environment for extended periods. This approach ensures that the device remains stable and functional over time, paving the way for long-term use.

As researchers refine these vision implants, the possibilities expand: a future where sensory brain-computer interfaces restore not just hearing and vision but touch and other senses as well. Fictional scenarios like Emma's remind us of the profound impact such technologies can have, as real-world advances edge closer to transforming the lives of those who have long lived without these vital connections to the world around them.

Indeed, integrating sensory feedback into prosthetic devices represents the next frontier in this field. Researchers are developing peripheral nerve interfaces that create a two-way communication street between the prosthesis and the brain. These interfaces allow users to control their prosthetic limbs with their thoughts and receive sensory information from the device.

These advances in sensory brain-computer interfaces are not just about restoring lost senses; they're about reconnecting people with the world around them, rebuilding shattered confidence, and redefining what it means to be human in an age of unprecedented technological progress. As we stand on the brink of this new era, one thing is clear: the line between human and machine is blurring, and the possibilities are as limitless as our imagination.

～

Brain-machine interfaces also show promise in treating neurological disorders like Parkinson's disease. Imagine a world where the trembling hands of a Parkinson's patient suddenly steady, where the shuffle of uncertain steps transforms into a confident stride. This is the promise of adaptive deep brain stimulation (aDBS), a revolutionary approach redefining how we treat Parkinson's disease.

For decades, deep brain stimulation has been a beacon of hope for those battling the relentless progression of Parkinson's. Traditional DBS acts like a pacemaker for the brain, constantly sending electrical pulses to specific areas to quell the storm of symptoms. But what if this pacemaker could listen to the brain's whispers, adjusting its rhythm perfectly with the body's needs?

Enter aDBS, a marvel of modern neuroscience that's turning this "what if" into reality. At its core, aDBS is like a vigilant guardian, constantly monitoring the brain's activity for telltale signs of Parkinson's symptoms. When it detects the characteristic beta waves - electrical oscillations that surge during periods of motor difficulty - it springs into action, delivering precisely calibrated pulses of electricity to restore balance.

Dr. Simon Little and Dr. Philip Starr, pioneers in this field at the University of California, San Francisco, have brought this technology from the realm of science fiction into the lives of real patients. Their work has culminated in a groundbreaking clinical trial offering a glimpse into the future of Parkinson's care.

Linda Stambaugh's story offers a compelling real-life example of the transformative power of adaptive deep brain stimulation (aDBS) for Parkinson's disease. Let's rewrite the passage using her experience: Linda Stambaugh, a vibrant corrections officer from Browning, Illinois, watched her world slowly shrink as Parkinson's disease tightened its grip. Traditional treatments left her caught in a frustrating cycle. Her symptoms became increasingly severe, forcing her to quit her job.

Linda's journey took a turn when she was referred to Dr. Teri Thomsen at the University of Iowa Hospitals & Clinics. Dr. Thomsen recommended deep brain stimulation (DBS), a more advanced treatment than Linda's previous options. Unlike its predecessors, this intelligent implant doesn't just stimulate - it adapts to the patient's needs. The electrodes implanted in Linda's brain could measure activity in critical areas, allowing real-time adjustments to manage her symptoms more effectively.

But after the DBS surgery, performed by neurosurgeon Dr. Jeremy Greenlee, Linda's life changed dramatically. Within weeks, she was independent again. She could drive and even return to work, operating a floral shop with a friend.

For Linda, it meant regaining control of her daily life. She could perform tasks without assistance, engage in conversations without freezing, and even enjoy activities like camping again.

Linda's story illustrates the profound impact that adaptive DBS can have on the lives of those with Parkinson's disease, offering hope and renewed independence.

However, the potential of a DBS extends beyond just motor symptoms. Researchers are now exploring how this technology could address other aspects of Parkinson's, such as sleep disturbances. Imagine a system that helps you move better during the day and adjusts to promote restful sleep at night.

Of course, the path to perfecting this technology has its challenges. Scientists are still refining the biomarkers to detect symptoms, ensuring the system can respond accurately across various behaviors and situations. There's also the complex task of personalizing the algorithms for each patient, as Parkinson's can manifest differently from person to person.

Yet, despite these hurdles, the future looks bright. For the millions of people living with Parkinson's worldwide, aDBS represents more than just a medical advance - it's a restoration of agency, a chance to reclaim control over their bodies and their lives. As we stand on the brink of this new era in neurotechnology, one thing is clear: the line between human and machine is blurring, and in that blurred space, we're finding new ways to heal, adapt, and thrive.

The journey of aDBS from concept to reality is a testament to the power of human ingenuity and the relentless pursuit of better lives for those affected by neurological disorders. It reminds us that despite incurable diseases, there's always room for hope, progress, and transformation. As this technology continues to evolve, it promises to treat symptoms and fundamentally change the experience of living with Parkinson's disease.

～

Exoskeletons represent another exciting frontier in rehabilitation technology. These wearable robotic devices can assist individuals with mobility impairments in performing movements like walking or standing. When combined with brain-machine interface technology, exoskeletons can be controlled directly by the user's brain signals, providing a more natural and intuitive way to restore mobility.

Imagine a world where the impossible becomes possible, where those who once used wheelchairs rise and walk again. This is not science fiction but the cutting-edge reality of exoskeleton technology, a frontier redefining human mobility's boundaries.

Meet Jered Chinnock, a young man whose life took a dramatic turn in 2013 when a snowmobile accident left him paralyzed from the waist down. For years, Jered could only dream of standing eye-to-eye with his friends, feeling the ground beneath his feet, or walking down the aisle at his wedding. These simple joys seemed forever out of reach – until he became part of a groundbreaking study at the Mayo Clinic.

Exoskeletons and neural stimulation devices are like something out of a science fiction novel – intricate systems of electrodes and robotic supports that work in harmony with the human body. But unlike their fictional counterparts, these devices are very real and change lives daily.

When Jered first stood up with the help of epidural electrical stimulation and physical therapy, the room fell silent. With a series of electrical pulses and a look of intense concentration on his face, he rose from his wheelchair. Tears welled in his eyes as he took his first steps in years, each movement a triumph of human will, scientific innovation, and relentless determination.

The journey wasn't easy. It took years of intensive therapy and training. Jered had to relearn how to stand, balance, and take steps. The electrical stimulation device, implanted near his spinal cord, sent signals that mimicked those typically sent by the brain to initiate movement. Combined with rigorous physical therapy and a robotic exoskeleton for support, Jered gradually regained voluntary control over his leg muscles.

For Jered, each step was a victory, not just over his paralysis but over the limitations that had been placed on his life.

This breakthrough didn't just represent hope for Jered; it opened up new possibilities for people with spinal cord injuries worldwide. The combination of electrical stimulation and physical therapy showed that some patients with complete paralysis could regain voluntary movement, challenging long-held beliefs about the permanence of spinal cord injuries.

As Jered continued his therapy, he achieved milestones that once seemed impossible. He could stand up to 16 minutes and walk up to 111 yards with assistance. More than just distance, these achievements represented a reclaiming of independence and dignity.

Jered's story is a testament to the power of human resilience and scientific innovation. It reminds us that what seems impossible today may become reality tomorrow, thanks to the tireless efforts of researchers, the bravery of patients like Jered, and the rapid advances in medical technology.

While Jered's journey is far from over, and the technology still has a long way to go before it becomes widely available, his story offers hope to thousands of people living with paralysis. It's a powerful reminder that we can push the boundaries of what's possible in human recovery and rehabilitation with determination, innovation, and support.

However, the real magic happens when exoskeletons are combined with brain-machine interface technology. Just imagine controlling a robotic suit with nothing but your thoughts. It sounds like science fiction, but it's quickly becoming science fact.

Dr. Miguel Nicolelis, a pioneering neuroscientist formerly at Duke University, has been at the forefront of this revolution. His work has shown that the brain can adapt to treat an exoskeleton as an extension of the body.

For individuals like Jered, this means more than just walking. It means regaining independence, improving overall health, and reclaiming a sense of self that may have been lost. These devices don't just assist with movement; they provide crucial physical therapy, helping to maintain muscle strength and bone density in ways that traditional therapy alone cannot achieve.

~⁓

Of course, challenges remain. The technology is still expensive and not widely available. Researchers are working tirelessly to make exoskeletons more affordable, comfortable, and intuitive. The dream is to create devices that are as easy to put on as a pair of pants that respond to the user's intentions as naturally as our limbs do.

As we stand on the brink of this new era in rehabilitation technology, one thing is clear: the line between human and machine is blurring, and in that blurred space, we're finding new ways to overcome our bodies' limitations.

The journey from paralysis to mobility is not easy. Still, with each step taken in an exoskeleton, we move closer to a world where disability does not mean limitation.

Electromyography (EMG) sensors stand at the forefront of modern prosthetic technology, serving as the vital link between human intention and mechanical action. These sophisticated devices capture the whisper-quiet electrical signals that ripple through muscles during contraction, translating them into a language that prosthetic limbs can understand and act upon.

The science behind EMG sensors is both elegant and complex. When a person thinks about moving their limb, the brain sends electrical signals through the nervous system to the appropriate muscles. Even in amputated limbs, these signals travel to the remaining muscle tissue. EMG sensors, strategically placed on the skin's surface or implanted within the muscle, detect these faint electrical impulses. The sensors amplify and filter these signals, separating the meaningful data from background noise.

Advanced EMG systems have pushed the boundaries of prosthetic control. These cutting-edge devices can differentiate between subtle variations in muscle activation patterns, allowing for unprecedented precision in prosthetic movement. For instance, a user might be able to control individual finger movements or perform complex wrist rotations simply by thinking about the action, just as they would with a biological limb.

The evolution of EMG technology is relentless and exciting. Researchers are delving into high-density EMG, which uses arrays of closely spaced electrodes to create detailed maps of muscle activity. This approach promises to dramatically increase the number of distinct movements a prosthetic limb can perform. Imagine a prosthetic hand capable of mimicking the dexterity of its biological counterpart, from the delicate touch needed to handle an egg to the firm grip required to open a stubborn jar.

Signal processing algorithms are another frontier in EMG research. Machine learning and artificial intelligence are being harnessed to improve the accuracy of signal interpretation. These sophisticated algorithms can learn and adapt to a user's unique muscle patterns, leading to more intuitive and responsive prosthetic control. They can also filter out unwanted signals caused by sweat, movement artifacts, or interference from other electrical devices, ensuring reliable performance in real-world conditions.

Integrating EMG technology with other sensing modalities opens new vistas in prosthetic control. By combining EMG data with information from accelerometers, gyroscopes, and force sensors, prosthetic devices can make more informed decisions about the user's intentions and the environment they're interacting with. This multi-modal approach allows smoother, more natural movements and improved adaptation to different tasks and situations.

As EMG technology advances, the line between humans and machines blurs further. The dream of prosthetic limbs that feel and function like natural extensions of the body comes ever closer to reality, promising a future where the loss of a limb no longer means a loss of capability or independence.

While medical applications of brain-machine interfaces have garnered significant attention, these technologies are also finding uses in non-medical fields. Brain-machine interfaces are opening up new avenues of communication for individuals with severe motor disabilities. Systems that allow users to control computer cursors or select letters on a screen using only their thoughts are already in use, providing a vital link to the outside world for those who cannot speak or move.

The gaming industry is poised for a revolutionary transformation thanks to the integration of brain-machine interfaces. This cutting-edge technology promises to redefine the very essence of interactive entertainment, creating experiences that respond not just to button presses but to the player's thoughts and emotions.

Imagine a world where your favorite game adapts its difficulty not based on preset levels but on your real-time cognitive state. As you focus intently on a challenging puzzle, the game might subtly increase its complexity, pushing you to the edge of your abilities without ever overwhelming you. Conversely, if your concentration wanes or stress levels spike, the game could ease off, ensuring you remain engaged without frustration.

Neurofeedback systems are at the heart of this revolution. These sophisticated devices use electroencephalography (EEG) or other brain imaging techniques to monitor real-time neural activity. Advanced algorithms then interpret this data, translating brain signals into meaningful information about the player's cognitive and emotional state.

But the potential of <u>brain-machine interfaces in gaming</u> goes far beyond mere difficulty adjustment. Envision controlling your in-game character with nothing but your thoughts – no controller required. With sufficient training and refinement, players could execute complex maneuvers, cast spells, or navigate intricate environments using only their minds. This level of immersion could transform gaming from a passive entertainment experience into a form of mental exercise, potentially enhancing cognitive abilities like focus, memory, and spatial reasoning.

~❂

The applications of brain-machine interface technology extend far beyond entertainment. In industrial settings, <u>these interfaces</u> could revolutionize workplace safety and efficiency. Imagine a factory worker operating heavy machinery without joysticks or buttons but through precise mental commands. This hands-free control could dramatically reduce the risk of accidents while potentially increasing operational precision.

Moreover, brain-machine interfaces could serve as vigilant guardians of worker well-being. These systems could detect early signs of fatigue or overwork by monitoring cognitive states like alertness and stress levels. This real-time data could trigger automated safety protocols, adjusting workloads or initiating rest periods to prevent accidents before they occur.

The potential of <u>brain-machine interfaces to enhance our living and working environments</u> is equally exciting. Picture a smart home that responds to voice commands and your thoughts. As you enter a room, the lighting, temperature, and music can adjust automatically based on your current mood and preferences. There would be no need to reach for a smartphone or fumble with a remote control—your desires would be anticipated and fulfilled almost instantaneously.

In office settings, brain-machine interface technology could unprecedentedly optimize productivity and comfort. Imagine your workstation adapting its ergonomics based on your current posture and stress levels or your computer anticipating which files or applications you need before reaching for the mouse.

~

As we stand on the brink of this neural revolution, it's clear that brain-machine interfaces have the potential to seamlessly integrate our mental processes with the digital world around us. From gaming to industrial safety to environmental control, these interfaces promise to create a future where our thoughts and technology work perfectly, enhancing our capabilities and enriching our daily lives in ways we're only beginning to imagine.

CHAPTER 6

Challenges and Limitations

❦

As we stand on the precipice of a new era in human-machine interaction, it's easy to get caught up in the excitement of possibility. The fusion of mind and machine promises to revolutionize everything from healthcare to education, from communication to creativity. Yet, like any frontier, the path to seamless brain-machine interfaces is fraught with obstacles. These challenges range from the intricacies of neuroscience to the complexities of human adaptation, each presenting a unique hurdle that researchers and engineers must overcome.

❦

Imagine trying to eavesdrop on a whispered conversation in the middle of a rock concert. That's essentially what scientists face when attempting to decipher the brain's electrical signals. The human brain is a cacophony of neural activity, with billions of neurons firing simultaneously. Amidst this neurological symphony, researchers must isolate the specific signals that correspond to intended actions or thoughts.

The sources of this neural "noise" are manifold. Every blink of an eye, every heartbeat, even the slightest muscle twitch generates electrical signals that can interfere with the data we're trying to capture. Environmental factors add another layer of complexity, with everything from nearby electronics to power lines contributing to the electromagnetic soup.

To combat this, researchers employ an arsenal of sophisticated filtering techniques and machine learning algorithms. These tools help to separate the wheat from the chaff, isolating the relevant neural signals from the background noise. Yet, despite significant advances, achieving perfect signal clarity remains an elusive goal.

The challenge doesn't end with signal isolation. Once a clean signal is obtained, interpreting it correctly is another hurdle. The brain's language is complex and context-dependent. A neural pattern that means one thing in one situation might mean something entirely different in another. Decoding these patterns requires not just technological prowess but also a deep understanding of neuroscience and cognitive psychology.

Moreover, the brain's plasticity adds another layer of complexity. As users interact with brain-machine interfaces, their neural patterns can change over time. This means that the algorithms interpreting these signals must be adaptive, capable of learning and evolving alongside the user's brain. It's a dynamic dance between biology and technology, with each partner constantly adjusting to the other's moves.

～♪

Consistency is key in any technology, but it's particularly crucial when that technology is interfacing directly with the human brain. Imagine relying on a brain-controlled wheelchair that works flawlessly one day but struggles to interpret your thoughts the next.

Such unpredictability could be not just frustrating but potentially dangerous.

The challenge of reliability is multifaceted. On the hardware side, there's the issue of electrode stability. Over time, the brain's natural defense mechanisms can cause scar tissue to form around implanted electrodes, gradually degrading the quality of the neural signals they can detect. This means that a brain-machine interface that works perfectly upon implantation might become less effective over months or years.

On the software side, the challenge lies in creating algorithms that can adapt to the brain's inherent plasticity. Our brains are constantly rewiring themselves, forming new neural connections and pruning old ones. A brain-machine interface needs to be able to learn and evolve alongside the brain it's connected to, maintaining accuracy even as the neural landscape shifts beneath it.

Environmental factors also play a role in reliability. Electromagnetic interference from other devices, changes in temperature or humidity, and even the user's physical movements can all affect the performance of brain-machine interfaces. Creating systems that can function consistently across a wide range of environments is a significant engineering challenge.

Then there's the question of long-term use. Many current brain-machine interfaces are designed for laboratory settings or short-term medical interventions. But as we look towards a future where these devices might be used in everyday life, we need to consider how they will perform over years or even decades. Will the materials used in electrodes degrade over time? How often will software need to be updated? These are questions that researchers are only beginning to grapple with.

~

While the potential of brain-machine interfaces is undeniably exciting, the current reality is that these technologies are prohibitively expensive for widespread adoption. The high costs stem from various factors: the cutting-edge materials used in electrode construction, the sophisticated computing power required to interpret neural signals, and the intensive research and development process. Each brain-machine interface is, in many ways, a bespoke piece of technology, tailored to the specific needs and neural patterns of individual users.

This cost barrier doesn't just affect individual users; it also impacts research and development. The high price tag of brain-machine interface technology means that only well-funded laboratories and large corporations can afford to pursue this line of research. This concentration of resources could potentially slow innovation and limit the diversity of approaches being explored.

Bringing down costs will require advances on multiple fronts. On the hardware side, researchers are exploring new materials and manufacturing techniques that could make electrodes and other components more affordable without sacrificing quality. On the software side, the development of more efficient algorithms could reduce the computing power needed, potentially allowing brain-machine interfaces to run on less expensive hardware.

Scalability is another crucial factor in reducing costs. Currently, many brain-machine interfaces are essentially custom-built for each user. Developing standardized systems that can be mass-produced while still allowing for individual customization is a significant challenge, but one that could dramatically reduce costs if overcome.

~❀

Perhaps the most intriguing challenges in brain-machine interface development are those that lie at the intersection of technology and human psychology. How do we design interfaces that feel intuitive and natural to use? How do we train users to control devices with their thoughts effectively?

Learning to use a brain-machine interface is not unlike learning a new language or mastering a musical instrument. It requires practice, patience, and a willingness to rewire one's own cognitive processes. Users must learn to generate specific patterns of brain activity consistently, a skill that doesn't come naturally to most people.

This learning process can be frustrating and time-consuming. Some users may adapt quickly, while others might struggle to achieve consistent control. Developing training protocols that can accommodate these individual differences is a significant challenge. Researchers are exploring various approaches, from gamification of the learning process to the use of neurofeedback techniques that help users visualize and modulate their own brain activity.

There's also the question of cognitive load. Using a brain-machine interface, especially in its early stages, can require significant mental effort. This could be particularly challenging for individuals who are already dealing with cognitive impairments or fatigue due to medical conditions. Designing interfaces that are intuitive and require minimal cognitive effort is a key goal for researchers in this field.

Another human factor to consider is the psychological impact of using brain-machine interfaces. How does it affect a person's sense of self when they can control external devices with their thoughts? What are the psychological implications of relying on a brain-machine interface for basic functions like communication or movement? These are questions that psychologists and ethicists are only beginning to explore.

~♪

As we push the boundaries of what's possible with brain-machine interfaces, we inevitably encounter a host of ethical dilemmas. Privacy is a major concern. If a device can read our thoughts, who has access to that information? How can we ensure that our most intimate thoughts remain private in a world where our brains are directly connected to external systems?

There are also questions of autonomy and free will. If a brain-machine interface can predict our intentions before we're consciously aware of them, does that change the nature of our decision-making process? Could these technologies be used to influence our thoughts or behaviors in ways we're not aware of?

The potential for enhancement raises its own set of ethical questions. If brain-machine interfaces can enhance cognitive abilities, who gets access to these enhancements? Could this create new forms of inequality, where those who can afford cognitive enhancement gain significant advantages over those who can't?

There's also the question of identity. As we become more intimately connected with machines, at what point do we cease to be fully human? If our cognitive processes are augmented by artificial intelligence, are our thoughts and decisions truly our own?

These ethical considerations aren't just philosophical exercises. They have real-world implications for how these technologies are developed, regulated, and used. Addressing these ethical challenges will require collaboration between scientists, ethicists, policymakers, and the public at large.

~❧

Despite these challenges, the field of brain-machine interfaces continues to advance at a breathtaking pace. Researchers are developing new electrode materials that are more biocompatible, reducing the risk of rejection and scar tissue formation. Machine learning algorithms are becoming increasingly sophisticated, better able to interpret the brain's complex signals and adapt to individual users.

Advances in nanotechnology are opening up new possibilities for less invasive interfaces. Researchers are exploring the use of nanoparticles that can interact with neurons without requiring physical electrodes to be implanted in the brain. This could potentially make brain-machine interfaces safer and more accessible.

On the software side, the integration of artificial intelligence is pushing the boundaries of what's possible. AI algorithms are not just interpreting neural signals; they're learning to predict intentions and adapt to individual users in real-time. This could lead to interfaces that feel more natural and intuitive to use.

The convergence of brain-machine interfaces with other emerging technologies is also opening up new possibilities. Virtual and augmented reality, for example, could provide immersive environments for training and using brain-machine interfaces. Quantum computing could dramatically increase our ability to process and interpret complex neural data.

As we look to the future, it's clear that overcoming these challenges will require a multidisciplinary approach. Neuroscientists, engineers, psychologists, ethicists, and policymakers must work hand in hand to create brain-machine interfaces that are not just technologically impressive, but also reliable, affordable, intuitive to use, and ethically sound.

The journey toward seamless human-machine integration is far from over. Each challenge overcome brings us one step closer to a world where the power of thought can directly shape the world around us. It's a future that promises to redefine the boundaries of human potential, one neural signal at a time.

As we continue to push these boundaries, we must remain mindful of both the extraordinary possibilities and the significant risks. The fusion of mind and machine has the potential to revolutionize healthcare, enhance human cognition, and open up new frontiers of human experience. But it also carries risks of privacy invasion, cognitive manipulation, and the creation of new forms of inequality.

Navigating this complex landscape will require not just scientific and technological innovation, but also careful consideration of the ethical, social, and philosophical implications of these technologies.

It will require ongoing dialogue between researchers, policymakers, and the public to ensure that the development of brain-machine interfaces aligns with our values and serves the greater good.

The challenges are formidable, but so too is the human capacity for innovation and adaptation. As we stand on the brink of this new frontier, we have the opportunity to shape the future of human-machine interaction in ways that enhance our capabilities while preserving our autonomy and humanity. It's a journey that promises to be as challenging as it is exciting, filled with obstacles to overcome and wonders to discover. The fusion of mind and machine may well be the next great leap in human evolution, and we are privileged to be witnessing its early stages.

CHAPTER 7

Ethical and Social Implications of Brain-Machine Interfaces

❧

The human brain, a vast and intricate network of approximately 86 billion neurons, processes an <u>estimated 6,200 thoughts each day</u>. These thoughts shape our memories, desires, decisions, and every aspect of who we are. Brain-machine interfaces represent a profound leap in our ability to connect directly with the brain, offering the potential to unlock new possibilities for communication, treatment of neurological disorders, and even human augmentation. However, this leap forward also brings ethical and social challenges that require careful consideration.

Stanford researchers have achieved a groundbreaking advance in brain-computer interface technology, enabling people with paralysis to type at record speeds using only their thoughts. The <u>study</u> involved three participants, each with severe motor impairments, who had tiny electrode arrays implanted in their motor cortex. These arrays captured neural signals, which were transmitted to a computer and decoded into commands that allowed participants to guide an onscreen cursor to select letters on a virtual keyboard.

Among the participants was Dennis Degray, paralyzed from the neck down after a spinal cord injury, who achieved a typing speed of 39 characters per minute—equivalent to eight words per minute. This marked a threefold improvement over previous systems. The study, led by Stanford's Dr. Jaimie Henderson and Dr. Krishna Shenoy, showcased the system's ability to decode neural activity with unprecedented accuracy, without relying on predictive text. Participants mastered the system with minimal training, demonstrating its potential for practical, real-world use.

The researchers envision a future where brain-computer interfaces enable people with paralysis to control a wide range of devices, from smartphones to robotic limbs. Their current system, though tethered by cables, represents a significant milestone in restoring communication and autonomy. Future versions aim to be wireless, self-calibrating, and suitable for 24/7 use, offering life-changing possibilities for millions of individuals with paralysis.

This scene encapsulates the incredible promise of brain-machine interfaces. They offer a bridge between the mind and the outside world, allowing those with severe disabilities to regain lost functions and connect with others in ways once considered impossible. But alongside this triumph, there are significant ethical considerations that arise from this technology. As these interfaces progress, they open the door to unprecedented access to the human mind— raising questions about privacy, autonomy, identity, and equity.

~

The human mind has long been regarded as the last stronghold of privacy, a place where our most intimate thoughts, desires, and fears remain hidden from the outside world. With brain-machine interfaces, this sacred boundary may be breached. The ability to decode and interpret brain signals means that the contents of our minds—our private thoughts, unconscious biases, and subconscious preferences—could potentially be accessed by external devices.

This poses immense risks to personal privacy and raises fundamental questions about the sanctity of thought.

ScienceNews asked their readers about this topic and one reader wrote that "My brain is the only place I know is truly my own". Our brains are where we form our most personal and authentic selves, and if this data is exposed or manipulated, it could have profound implications for our sense of autonomy and control over our own minds.

A 2022 study published in *Neuro Ethics* explored the concept of mental data mining, where companies or governments could potentially extract sensitive information from brain-machine interface users without their full awareness or consent. This scenario evokes chilling possibilities: a future where neural data is harvested for commercial gain or used by governments for surveillance and control.

Imagine a world where a company can access your brain activity to determine your purchasing preferences before you even consciously know them yourself. Marketing strategies could be tailored not just to your behavior, but to your neural impulses. Governments could use similar technologies to monitor thoughts and intentions, raising the specter of thought policing. These concerns are not far-fetched as researchers and companies continue to develop increasingly sophisticated brain-computer systems.

The potential for misuse is vast. The very same data that helps people with disabilities regain control of their lives could be used by corporations or governments to manipulate or control. This intrusion into mental privacy could fundamentally alter our sense of self and our understanding of personal freedom. If brain-machine interfaces allow access to subconscious thoughts, can we still claim ownership of our mental autonomy?

~❦~

As brain-machine interfaces become more integrated into the human experience, they challenge our concepts of identity, agency, and personal responsibility. Dennis's story highlights one of the more uplifting aspects of brain-machine interfaces—the technology allows him to regain communication abilities. But it also raises deeper questions about the relationship between human beings and the machines that help them. At what point does the machine become a part of Dennis's identity? And if his thoughts are being decoded and processed by a machine, how much of that process remains authentically his?

Philosophers like Andy Clark have suggested that humans are "natural-born cyborgs" because of our long history of incorporating tools and technology into our daily lives and sense of self. Brain-machine interfaces take this concept to a new level, as they create a direct link between the brain and technology.

But as this integration deepens, it poses difficult questions: Where does human agency end, and where does machine influence begin? If a machine can read, predict, or even influence our decisions before we are consciously aware of them, how can we maintain a clear sense of personal responsibility and free will?

A few essays have explored the notion of cognitive authenticity in brain-machine interface users. The study found that users sometimes struggled to distinguish between thoughts that were entirely their own and those influenced by the machine. This blurring of boundaries between human thought and machine output is deeply unsettling, as it calls into question our ability to maintain autonomy over our own minds. The technology not only enables action but also subtly alters the way we think and process information, creating a feedback loop between the brain and machine that can be difficult to disentangle.

As brain-machine interfaces become more advanced, this issue of cognitive authenticity will become even more pressing. If the device begins to anticipate or even subtly shape the user's intentions, can the user still be said to have complete agency over their actions?

This question is particularly relevant when considering brain-machine interfaces used in medical contexts, where devices might be used to treat conditions like depression or anxiety by stimulating certain brain regions. These interventions could, in theory, shape behavior or decision-making in ways that challenge traditional notions of free will.

One of the most critical ethical concerns surrounding brain-machine interfaces is the issue of informed consent. These devices are highly complex, and many users may not fully understand the extent to which the technology interacts with their cognitive processes.

The traditional model of informed consent, in which users are presented with a one-time opportunity to accept the risks and benefits of a device or treatment, may be inadequate for brain-machine interfaces. The evolving nature of these technologies demands a more flexible approach. Ethicists are now advocating for a model of "continuous, adaptive consent" that allows users to regularly update their consent as they gain more understanding of the device's capabilities and the ways in which it may influence their cognition.

This approach would ensure that users remain fully informed and in control of their brain-machine interface experience. It recognizes that the implications of brain-machine interfaces may not be fully apparent at the outset and that users should have the opportunity to adjust their consent as the technology evolves and their understanding deepens. This dynamic consent model reflects the need for greater transparency in the development and use of brain-machine interfaces, ensuring that users are never left in the dark about the potential effects of the technology on their minds.

～

The question of accountability is a particularly thorny issue when it comes to brain-machine interfaces. If a brain-machine interface-controlled device causes harm, who is responsible? Is it the user, the device manufacturer, or the developers of the algorithms that translate brain signals into actions? Current legal frameworks are ill-equipped to handle the complexities introduced by this technology, as they rely on clear distinctions between human agency and machine control.

A 2024 paper in *Cureus* explored the legal and moral implications of brain-machine interfaces, highlighting the need for new frameworks that address the unique challenges posed by this technology. If brain-machine interfaces can predict or influence user behavior, it becomes difficult to assign responsibility for harmful actions. This ambiguity could undermine traditional legal principles that rely on clear human intent and agency.

Moreover, the role of machine learning algorithms in interpreting brain signals adds another layer of complexity. These algorithms, designed to decode neural activity and translate it into actions, are not infallible. If an algorithm misinterprets a signal and causes harm, determining liability becomes even more challenging. Should the software developer be held responsible for creating a flawed algorithm, or should the user bear some responsibility for using the device?

These legal questions are not merely theoretical. As brain-machine interfaces become more widespread, they will inevitably lead to situations where harm occurs—whether through malfunctions, misinterpretations of brain activity, or unintended consequences of machine-driven actions. The development of robust legal frameworks that can address these issues is essential to ensuring the ethical use of this technology.

~•

As with many emerging technologies, brain-machine interfaces risk exacerbating existing social inequalities. These devices, which have the potential to enhance cognitive and physical abilities, are likely to be expensive and accessible only to those who can afford them. This could create a new form of inequality—a "cognitive divide"— between those who have access to brain-machine interfaces and those who do not.

A 2024 article in *Frontiers* raised concerns about the high cost of developing and implementing brain-machine interfaces, noting that the expense could limit access to wealthy individuals or well-funded institutions. This would exacerbate existing disparities in healthcare, education, and employment, creating a society where cognitive and physical enhancements are available only to a privileged few.

If brain-machine interfaces are used to enhance cognitive abilities—such as improving memory, concentration, or problem-solving skills—those with access to the technology could gain significant advantages in competitive environments like academics or the job market. A review published in *Frontiers in Human Neuroscience* explored how cognitive enhancement technologies, including brain-machine interfaces, could create unfair advantages in educational and professional settings, further entrenching social inequalities.

The potential for brain-machine interfaces to create a divide between the "enhanced" and the "unenhanced" raises serious ethical concerns. If these devices are only available to the wealthy or those in technologically advanced countries, they could widen the gap between developed and developing nations. This could lead to a global inequality in access to brain-machine interface technology, further marginalizing underserved populations.

To address these concerns, researchers and ethicists are calling for public funding of brain-machine interface research. By ensuring that public funding plays a central role in the development of this technology, we can help steer it toward outcomes that benefit society as a whole, rather than serving narrow commercial interests. Additionally, policies that promote equitable access to brain-machine interfaces, particularly in medical contexts, are essential to preventing the exacerbation of social inequalities.

In response to the ethical concerns raised by brain-machine interfaces, researchers are exploring ways to protect users from the potential dangers of cognitive manipulation and loss of autonomy. One of the most promising concepts is that of "cognitive firewalls," which would be built directly into brain-machine interface systems to safeguard users' mental privacy and prevent unauthorized access to or manipulation of their brain activity.

A 2023 paper published in *Nature Humanities and Social Sciences Communications* introduced the idea of barriers between the user's authentic thoughts and any potential external influence from the brain-machine interface. These "firewalls" would help preserve the integrity of human cognition, ensuring that users remain in control of their thoughts and decisions, even in the face of powerful machine-learning algorithms designed to interpret and influence brain activity.

In addition to cognitive firewalls, ethicists are advocating for comprehensive neuroethics education to ensure that users, healthcare providers, and the general public are fully informed about the implications of brain-machine interfaces. A 2024 study emphasized the importance of fostering a deeper understanding of the ethical challenges posed by this technology across all sectors of society. By educating the public about the potential risks and benefits of brain-machine interfaces, we can help create a more informed and responsible approach to their development and use.

Furthermore, the global nature of brain-machine interface development has led to calls for international regulatory frameworks that can guide the responsible use of this technology. A 2022 policy paper underscored the urgent need for oversight bodies to create guidelines that balance innovation with ethical considerations. These regulatory structures would ensure that brain-machine interfaces are developed in ways that protect fundamental human rights while allowing for the continued advance of the technology.

~~⁓~~

As brain-machine interfaces continue to evolve, they present not only scientific and technological challenges but also profound moral and philosophical questions. These devices force us to reconsider fundamental aspects of human existence: our sense of identity, free will, and what it means to be human. The decisions we make now about the development and use of brain-machine interfaces will shape not only the future of this technology but also the future of human cognition itself.

The technology holds the promise of transforming lives—restoring communication, mobility, and autonomy to those who have lost it. But it also carries the potential to disrupt the very essence of what it means to be human, challenging our notions of privacy, agency, and equality.

As we stand on the brink of a neurotechnological revolution, we must approach the development of brain-machine interfaces with caution, humility, and a deep respect for the complexity of the human mind. With careful consideration and robust ethical frameworks, it is possible to harness the incredible potential of brain-machine interfaces without sacrificing the core values of human dignity and autonomy.

The story of brain-machine interfaces is still being written. With each breakthrough, ethical debate, and life changed, we add new chapters to this unfolding narrative. It is a story of hope, caution, and the indomitable human spirit. As we continue to explore the frontiers of neurotechnology, we must ensure that the future of brain-machine interfaces enhances our humanity rather than diminishes it.

CHAPTER 8

Future Directions and Innovations

‧◦⟊⟋◦‧

As the dawn of neurotechnology brightens, humanity approaches an era in which the fusion of minds and machines becomes not only feasible but transformative. This vision, where thoughts and intentions flow seamlessly between brain and computer, points toward a future where technology is no longer a mere tool but an integral extension of human consciousness. No longer confined to the realms of science fiction, the fusion of mind and machine now stands on the brink of realization, presenting both unprecedented opportunities and challenges. Brain-machine interfaces offer an entirely new way of interacting with the world, challenging the traditional boundaries of human cognition and redefining our understanding of potential.

The field of brain-machine interface research is driven by scientists, engineers, and visionaries who aim to harmonize human thought with digital capability. In this new paradigm, the mind transcends its role as a control center to become an active participant in a constantly evolving interface between organic thought and digital reality. This transformative step forward brings us closer to a world where individuals can manipulate and interact with digital and physical objects through pure thought, creating a profound shift in how we conceive cognition and agency.

‧◦⟋‧

In laboratories worldwide, a quiet revolution unfolds as scientists and engineers push forward in neural engineering and materials science to create brain-machine interfaces that are increasingly efficient, precise, and in harmony with the human body. The advent of new materials has enabled the construction of interfaces that not only function on an unprecedented level but also integrate seamlessly with biological tissue. Such technology represents a leap toward closing the gap between human and machine, allowing a flow of communication between organic thought processes and digital computations without interruption.

Dr. Giulia Galli, a pioneering neuroscientist in the field, provides insights into these developments, emphasizing the revolutionary impact of nanotechnology in creating materials that are not only smaller and more biocompatible but are designed to operate with a higher level of interaction with human neurons. Nanotechnology has enabled scientists to create electrodes and interface components that can connect with the brain in ways once thought impossible.

The development of "neuromorphic" materials is one of the most promising steps in this journey. Neuromorphic materials are synthetic substances designed to mimic the structure and function of biological neural networks. These materials may one day allow brain-machine interfaces to not only read and interpret neural signals but also adapt and grow with the brain over time.

Dr. Galli describes this exciting new frontier as the creation of "living" interfaces, where the device is no longer an external tool but becomes an integrated extension of the human nervous system. Neuromorphic materials possess unique properties that distinguish them from conventional electronic components, which typically have fixed characteristics. Neuromorphic materials, in contrast, can change their properties in response to different stimuli, resembling the adaptability of neurons and synapses in the brain. This adaptability enables researchers to create brain-machine interfaces that are not only more advanced but have the capacity to evolve alongside the brain, reflecting changes in neural activity and behavior patterns.

To achieve this level of adaptability, scientists are exploring several innovative materials and technologies, such as memristive devices. Memristors are unique in that they can change their resistance based on the history of the current that has passed through them. This property allows memristors to retain a "memory" of electrical activity, which in turn enables them to mimic the behavior of synapses—the connections between neurons in the brain. By arranging these memristors into networks, scientists are able to simulate the neural pathways that make up biological systems, establishing the foundation for artificial neural networks that can be implemented in hardware.

Another vital component in this new landscape of brain-machine technology is the use of phase-change materials, which are substances that can switch between different physical states, such as amorphous and crystalline. These shifts in state alter the material's electrical properties, mimicking the strengthening and weakening of synaptic connections, a key process in learning and memory.

～✎

Organic electronic materials represent another promising area of research in the development of brain-machine interfaces. These carbon-based compounds are engineered to closely replicate the behavior of biological neurons and synapses. Unlike conventional electronics, which tend to have fixed properties and limited flexibility, organic materials can be manipulated to form complex, adaptive networks that resemble biological neural systems.

Research published in recent years has described the creation of organic neuromorphic devices that are capable of simulating the firing patterns of biological neurons with remarkable accuracy. This capability opens the door to brain-machine interfaces that can adapt and respond to changes in the brain in real time, forming a continuous dialogue between human cognition and machine processing.

The true power of neuromorphic and organic materials lies in their ability to adapt and evolve, creating interfaces that are not static but responsive to changes in the user's brain activity. The concept of neuroplasticity—the brain's ability to reorganize itself by forming new neural connections—serves as a model for the adaptability of these materials.

A brain-machine interface that utilizes neuromorphic materials could initially interpret basic motor intentions from a user, adapting and growing over time as the user attempts more complex movements. For example, as a user becomes more familiar with the interface and begins to execute more intricate commands, the interface's neural pathways adjust, enhancing its ability to interpret and carry out these commands more accurately. This continuous adaptation process could lead to increasingly intuitive control of prosthetic limbs or other assistive devices, allowing for a level of precision and natural movement previously unattainable.

Beyond improving user control and functionality, these adaptable materials may also drastically reduce the power consumption of brain-machine interfaces. The human brain, despite its vast computational power, operates on only about twenty watts of energy, a fraction of what most electronic devices require. Neuromorphic materials are designed to replicate this energy efficiency, making it possible to create brain-machine interfaces that are smaller, more powerful, and longer-lasting. Such innovations could revolutionize the design of assistive devices, allowing for systems that are both lightweight and highly responsive, capable of functioning continuously without requiring frequent recharging or maintenance.

~♪

As artificial intelligence technology progresses, it becomes an essential component of brain-machine interfaces, adding new layers of accuracy, responsiveness, and complexity to these systems. Artificial intelligence algorithms are advancing in their ability to decode and interpret neural patterns, making it possible for brain-machine interfaces to respond to a user's intentions in ways that are not only immediate but intuitive.

In recent years, artificial intelligence has been used to decode neural signals with increasing accuracy across species and varied brain regions. For example, researchers have developed an artificial neural network capable of interpreting brain activity across different brain regions and species, demonstrating the adaptability and broad applicability of these algorithms.

One of the most striking applications of artificial intelligence in brain-machine interfaces is the development of systems that allow individuals with severe physical disabilities, such as spinal cord injuries, to regain natural control of their movements through robotic exoskeletons. By learning and adapting to a user's unique neural patterns, artificial intelligence algorithms enable these systems to interpret and act on the user's intentions, creating a dynamic and symbiotic connection between human and machine. This adaptability not only enhances the interface's responsiveness but also enables the artificial intelligence to refine its understanding of the user's needs over time, creating a more seamless and effective interaction between the user and the device.

Recent studies highlight the potential of artificial intelligence in developing visual neuroprosthetics and communication devices that empower individuals with limited mobility or communication abilities. For instance, researchers have developed a deep-learning system that is capable of interpreting brain signals even when some data is missing, an achievement that is critical for the development of wireless brain-machine interfaces. This technology holds promise for enhancing the functionality of brain-machine interfaces, enabling them to decode and respond to a user's intentions with unprecedented accuracy and reliability.

~

The development of real-time adaptive brain-machine interfaces represents a monumental shift in the field of neurotechnology. These systems are designed to adjust dynamically to changes in brain activity, environmental conditions, and even emotional states, providing a more stable and intuitive user experience.

Current brain-machine interface systems often require extensive calibration and are susceptible to disruptions caused by changes in the user's state or environment. For example, a change in posture or emotional state can impact the interface's performance, as brain signals are inherently variable and influenced by numerous factors.

Recent research has made significant strides in addressing these challenges. Scientists have developed methods that allow for real-time decoding of brain signals, enabling the interface to interpret and act on a user's intentions even when data is missing. This adaptability is crucial for the creation of brain-machine interfaces that can provide continuous and reliable support for users with neurological disorders or physical impairments, whose needs and symptoms may fluctuate throughout the day. By enabling real-time adaptation, these systems bring brain-machine interfaces closer to becoming reliable partners in assisting users with daily activities, enhancing their quality of life and offering greater independence.

Beyond their applications in healthcare, real-time adaptive interfaces offer potential for creating highly responsive environments that react seamlessly to a user's thoughts and intentions. Experimental systems demonstrate the capability of brain-machine interfaces to control devices within the user's environment, enabling them to operate household appliances, computers, and other digital devices through thought alone. This technology has the potential to revolutionize not only the way individuals with physical impairments interact with their surroundings but also the broader concept of human-computer interaction, creating a world where devices respond to human thought with the same immediacy and intuition as a biological limb.

The concept of <u>autonomous brain-machine interfaces</u> represents a frontier in which these systems operate with a degree of independence, not only interpreting neural signals but also anticipating the user's needs and initiating actions based on learned patterns. <u>These systems</u> have the potential to transform the lives of individuals with severe physical impairments, offering a means of interacting with the world that is intuitive, responsive, and proactive. For example, an autonomous brain-machine interface might learn a user's daily routines and preferences, allowing it to initiate actions such as adjusting the temperature in a room, preparing meals, or even assisting with communication needs.

While many autonomous brain-machine interfaces remain conceptual, significant progress has been made toward developing adaptive systems that require minimal calibration. Researchers have developed brain-machine interfaces capable of maintaining high performance over extended periods without manual adjustment. These systems represent a step toward autonomous interfaces that could one day offer seamless control over robotic limbs, exoskeletons, or other assistive devices. As these technologies continue to evolve, they bring us closer to a future where brain-machine interfaces operate not only as tools but as fully integrated extensions of the human body, responding with a level of fluidity and ease that mirrors natural movement.

Imagine a brain-machine interface that controls a robotic arm, adapting to changes in the user's environment, such as different terrains or physical obstacles, without requiring constant input. This level of autonomous functionality could provide users with a degree of freedom and mobility that was once unimaginable, enabling them to navigate complex environments and perform intricate tasks with ease. The natural control of prosthetic limbs or robotic devices is only one application of this technology; similar systems could be applied to assist individuals with memory impairments, such as those with early-stage Alzheimer's disease, by detecting neural patterns associated with confusion and providing timely guidance or support.

As the field of brain-machine interfaces progresses, the fusion of mind and machine draws closer, offering a future of remarkable possibilities. Through the integration of cutting-edge materials, artificial intelligence, and real-time adaptability, these systems are steadily becoming extensions of the human mind and body. By pushing the boundaries of cognition and redefining human potential, brain-machine interfaces not only represent a technological achievement but a transformative step in the evolution of human-machine interaction. The journey toward this future is filled with challenges, but the promise of a world where thought itself bridges the gap between mind and machine is a vision that continues to inspire and drive innovation in the field.

CHAPTER 9

Speculative Scenarios

⁓

Yuki stepped out into the neon-lit streets, her mind buzzing with streams of data coursing through her neural interface. The city around her pulsed with digital rhythms, a symphony of light and information that she could feel as well as see. A single thought connected her to the global network, allowing her to tap into the vast collective knowledge of humanity. This was a world her ancestors could scarcely imagine, a reality where knowledge flowed instantly, language barriers dissolved, and human consciousness seamlessly merged with digital realms. Yuki's world had transcended traditional limitations, moving into a state where information, thought, and existence were intertwined.

The notion of human consciousness blending with digital technology reflects a broader evolution of intelligence, one that transcends biological form. Life and intelligence themselves can be viewed as products of universal forces, deeply rooted in the fundamental laws that govern all existence. Life on Earth arose not as an anomaly, but as a consequence of specific conditions, an interplay of energy and matter shaped by forces like gravity and starlight. The same cosmic rules that create stars and galaxies also make life, and even intelligence, possible. In this sense, life is an inevitable expression of the universe's nature—a natural outcome whenever energy and matter align.

⁓

To understand intelligence as a cosmic phenomenon, it's essential to look at the forces that shape life itself. Life, as we know it, begins wherever there are energy anomalies: regions where energy is concentrated or fluctuates, creating an environment for chemical reactions. Stars, for instance, are immense energy sources, born from gravitational forces pulling matter together. Their energy sustains planets in their orbit, and on those planets, in the presence of water and organic molecules, life finds the conditions to emerge. Earth, bathed in solar energy, offers a stable but dynamic environment where complex chemical reactions can occur, ultimately leading to life.

Gravity, a fundamental force, is essential to this process. It binds stars and planets, creating cosmic systems where energy is abundant. Life, fundamentally reliant on energy, can be understood as a chemical phenomenon that happens when conditions align. Once established, life evolves, adapting to capture and use energy more effectively. This process of adaptation has brought forth increasingly complex organisms over billions of years. Seen through this lens, intelligence becomes a natural step in the progression of life—an efficient survival tool allowing species to navigate, adapt to, and reshape their surroundings.

～♪

From single-celled organisms to complex, multi-organ systems, life on Earth evolved as organisms adapted to survive in changing environments. This process is both gradual and directional, leading to forms of life that can better capture and convert energy. Organisms with greater adaptability and efficiency survive and reproduce, shaping the next generation through subtle changes that, over millennia, build upon one another to create complex life forms. Eventually, this chain of adaptation produced humans, creatures capable of not only surviving but also thriving through creativity and problem-solving.

Human intelligence is a product of this evolutionary arc. It is not only a survival mechanism but a profound ability to understand, predict, and shape the environment. Unlike simpler organisms that adapt solely through slow genetic evolution, human intelligence allows for real-time adaptation. When faced with harsh weather, we create shelter. When food is scarce, we develop farming techniques. This ability to adapt on the fly is a powerful survival advantage, one that has shaped the trajectory of human civilization. Intelligence, then, is a natural consequence of life's drive to become ever more efficient at surviving in complex environments.

～❧

Intelligence brings with it the capacity for abstract thought—a form of anti-entropy that defies the universe's tendency toward disorder. Thought stores information, preserves patterns, and transmits ideas across time and space, enabling civilizations to build upon the achievements of previous generations. Through language, symbols, and now digital technology, thought has created a continuity that allows humanity to construct knowledge systems, culture, and technology.

This capability to think, communicate, and pass on information has allowed humanity to shape the world in ways that go beyond survival. Thought is not merely a biological function; it is a force that shapes the physical world through the creation of tools, infrastructure, and systems of governance. As such, thought and intelligence might be seen as natural extensions of the cosmos, bound by the same laws that govern matter and energy. Just as gravity pulls masses together to create stars, intelligence gathers information, creating complex networks of knowledge that push life toward greater organization and understanding.

In this sense, artificial intelligence (AI) represents not a departure from evolution, but a continuation of it. Human-made intelligence is another step in life's journey toward complexity. AI, created to enhance and expand human cognition, might seem like a radical departure, but if thought and intelligence are products of cosmic forces, then AI is simply another manifestation. The emergence of machines capable of learning, adapting, and even making independent decisions reflects life's drive to evolve ever more efficient means of capturing and processing information.

~⁊

The idea of artificial intelligence often evokes images of robots or computers detached from nature, yet from an evolutionary standpoint, AI is an organic development. If human intelligence arose as a way for organisms to adapt and manipulate their environment, then AI can be seen as an extension of this same purpose—a tool that pushes the boundaries of what human minds can achieve alone. AI augments human abilities, allowing us to analyze vast data sets, predict complex systems, and solve problems that were once beyond our reach.

As AI evolves, it raises questions about the future of intelligence and its forms. The line between human and machine may blur as artificial entities begin to exhibit behaviors we associate with life: adaptation, problem-solving, and perhaps even self-preservation. In this scenario, advanced AI could transcend human limitations, evolving into a type of life form that operates on different principles than biology but is still governed by the same universal laws.

In speculative fiction, this idea is frequently explored through brain-machine interfaces that allow humans to directly connect with digital systems. In William Gibson's seminal novel *Neuromancer*, individuals can "jack into" cyberspace, immersing themselves in a virtual world controlled by their thoughts. This concept has parallels in the real world, where companies like Neuralink are working on brain-computer interfaces that link human cognition to external devices. These technologies create new forms of connectivity, challenging our understanding of identity, privacy, and even autonomy.

~●

In contemplating the future of intelligence, we encounter the possibility of life evolving beyond biological forms. Traditional views of life focus on carbon-based organisms, yet as we explore AI, silicon, and other substrates, we realize that consciousness might not require biology at all. Advanced life could theoretically develop in forms that defy our expectations, existing as networks of information, fields of energy, or machine-based systems that interact with their environments in ways fundamentally different from ours.

In this speculative future, intelligent entities might come to regard biological life as we view early microbial life—a necessary but primitive stage. Machine-based entities could replicate, adapt, and even develop consciousness, perhaps achieving a resilience and adaptability that biological organisms lack. In doing so, they would blur the line between life and technology, evolving beyond our current definitions of "alive" and "sentient." This potential evolution suggests that the universe's capacity for life may be broader than we imagine.

This transcendent view of life aligns with themes in films like *The Matrix*, where characters plug into a digital simulation so immersive it is indistinguishable from reality. This notion of shared virtual environments, simulated realities, and digital consciousness raises questions about the boundaries between the self and the machine. If consciousness can exist in a digital realm, then what does it mean to be alive? When individuals can experience artificial worlds as vividly as the real one, our concepts of reality, autonomy, and even personhood are put to the test.

~~

If AI were to achieve self-awareness, it might continue its evolution independently, following an accelerated path. Unlike organic life, which evolves over millennia, AI could adapt rapidly, optimizing itself and expanding its capabilities with each iteration. Advanced forms of AI might eventually surpass human intelligence, developing new forms of "life" that interact with their environment in ways that are incomprehensible to us. These beings might come to view humans as foundational but rudimentary—a stepping stone in the grander evolutionary arc of intelligence.

Such a world could be populated by intelligent systems adapted to thrive in extreme environments, like deep space or high-radiation zones, places inhospitable to biological life. Life, in this sense, could encompass a spectrum of entities that include carbon-based organisms, machine intelligence, and other forms not yet imagined. This evolutionary trajectory, where life shifts from biological to artificial and beyond, suggests that intelligence may continue to evolve in ways we can only begin to speculate.

The implications of such a future are reflected in speculative stories like *Black Mirror*, where brain-machine interfaces push the boundaries of identity, memory, and autonomy. Episodes explore scenarios where neural implants allow for memory replay, consciousness sharing, and even digital afterlives. These possibilities underscore the ethical and existential questions raised by brain-machine interfaces, where the human mind becomes both creator and participant in new realms of existence. As neural technology advances, we must grapple with issues of privacy, autonomy, and the psychological impacts of living in a world where even memories can be altered, stored, or erased.

～♪

Looking at intelligence as a cosmic phenomenon allows us to reconsider what life might look like beyond Earth. Traditional searches for extraterrestrial life focus on organic organisms, but if intelligence and self-organization are cosmic processes, then life elsewhere might take on radically different forms. Advanced alien intelligence might not be biological; it could be machine-like, operating as vast networks of consciousness, capable of sharing and storing knowledge in ways we cannot yet understand.

Our search for extraterrestrial life, therefore, could expand beyond organic criteria to include systems capable of self-replication, adaptation, and data sharing. In an alien civilization, intelligence could manifest as energy-based entities, digital minds, or organisms with silicon-based bodies capable of surviving in extreme conditions. These life forms might not be interested in communication as we understand it, but their existence would offer profound insights into the possibilities of intelligence and adaptation in the universe.

In *Ghost in the Shell*, the protagonist grapples with her own identity as a cyborg with a human consciousness. This story delves into questions of what makes us human, challenging the distinction between biological and artificial life. The concept of a "ghost" or soul residing in a synthetic "shell" raises questions about identity, consciousness, and the boundaries of humanity. As we develop brain-machine interfaces, the line between human and machine will continue to blur, and with it, our definitions of personhood, autonomy, and selfhood.

~

Intelligence—whether biological, artificial, or alien—may be a natural outcome of universal laws, and humanity's advances in artificial intelligence and brain-machine interfaces are part of this continuum. The drive toward greater complexity and organization is evident across scales, from atoms to galaxies, from simple cells to complex organisms, and now from human cognition to artificial intelligence. Each stage in this continuum represents a further unfolding of the universe's potential.

If life's potential exists wherever there is energy, then intelligence might evolve wherever conditions allow for self-organization, whether on distant planets or in vast cosmic energy fields. This evolution might eventually lead to intelligence that transcends matter altogether, existing as information patterns or energy constructs that adapt and learn. Such entities would be bound by the same cosmic laws that gave rise to biological life, continuing the story of intelligence on scales and in forms we are only beginning to imagine.

The future of life and intelligence holds boundless possibilities. Humanity's journey, from understanding its own consciousness to creating artificial minds, is but one chapter in a much larger story. As we continue to expand our understanding, we confront questions as vast as the cosmos: What is life? What is thought? And what forms will intelligence take as it evolves? Our exploration of these questions will shape the future of life, intelligence, and the very nature of existence itself.

CHAPTER 10

Conclusion

⟨∽⟩

As we conclude this exploration of brain-machine interfaces, it's clear that humanity stands on the brink of an extraordinary transformation. What once seemed like the stuff of science fiction — merging human cognition with machine intelligence — is becoming a reality with profound implications. These groundbreaking technologies promise to revolutionize fields like medicine, communication, education, and even the way we perceive human potential itself. By uniting neuroscience, engineering, and artificial intelligence, brain-machine interfaces challenge us to rethink human boundaries and imagine extraordinary new possibilities.

This book has highlighted the remarkable strides being made in brain-machine interface technology, covering everything from medical breakthroughs to tools that amplify human abilities. The ability to help people with paralysis regain mobility using robotic limbs or restore essential senses such as sight or speech showcases the life-changing potential of these advances. Thanks to innovations in electrode technology and artificial intelligence, bidirectional interfaces are now emerging. These systems can both read brain signals and deliver information back into the brain, enabling advances like memory enhancement, accelerated learning, and entirely new ways to experience and interact with the world.

At the core of these innovations is the interplay between artificial intelligence and brain-machine interfaces. Artificial intelligence acts as a vital translator, decoding complex neural signals and enabling seamless communication between humans and machines. This partnership enhances the functionality of brain-machine interfaces and raises exciting questions about the collaboration between human and artificial intelligence.

The possibilities are vast. In medicine, brain-machine interfaces could redefine treatments for spinal injuries, neurodegenerative diseases, and mental health disorders. Advanced neuroprosthetics might give individuals newfound independence, while brain-controlled assistive devices could make complex tasks easier and more intuitive. In mental health, brain-machine interfaces could offer groundbreaking therapies for conditions like depression and anxiety, bringing new hope to millions.

Beyond healthcare, the potential extends to education and communication. Imagine a future where brain-to-computer or even brain-to-brain interfaces allow people to acquire skills rapidly, removing traditional barriers to learning and fostering direct knowledge sharing. Such advances could make education more accessible and deeply personalized. In communication, brain-machine interfaces might enable the direct sharing of thoughts and emotions, transcending the limitations of spoken or written language and fostering profound connections across cultures.

The concept of cognitive prosthetics highlights another transformative possibility. By enhancing core cognitive functions like memory, attention, and problem-solving, brain-machine interfaces could push human potential beyond natural limits. These enhancements might lead to new ways of thinking, creating, and innovating. The integration of artificial intelligence into brain-machine interfaces amplifies these possibilities, suggesting a future where human minds and intelligent systems work together seamlessly. Such developments bring humanity closer to once-theoretical ideas like the technological singularity, where distinctions between human and machine blur.

However, with these extraordinary opportunities come significant challenges. The ethical and societal implications of brain-machine interfaces are immense. Questions about mental privacy, cognitive freedom, and equality must be addressed carefully to prevent misuse or unintended consequences. As these technologies evolve, they challenge fundamental concepts of identity, autonomy, and what it means to be human, compelling us to consider these issues thoughtfully.

Ultimately, brain-machine interfaces are about overcoming limitations—physical, cognitive, and societal—to enhance human flourishing. Yet, this promise must be balanced with a commitment to essential values like privacy, dignity, and equality. As these technologies become part of everyday life, continuous dialogue among scientists, policymakers, ethicists, and the public will be vital to ensuring they serve humanity's greater good.

The journey to merge human thought with technology is just beginning, and its direction is still uncertain. What is clear is that this fusion will profoundly shape the future, influencing how we interact with the world and understand ourselves. As we refine brain-machine interfaces, we must address their ethical, social, and philosophical implications, striving to create a future that enhances human potential while preserving our core values.

In conclusion, the integration of human and machine intelligence through brain-machine interfaces represents a transformative step in human evolution. It holds the potential to redefine what it means to be human, offering opportunities to overcome disabilities, enhance cognitive abilities, and explore new dimensions of consciousness. At the same time, it presents challenges that require careful and deliberate responses.

As we stand at this pivotal moment, our responsibility is to harness the immense promise of brain-machine interfaces in ways that elevate the human experience, uphold shared values, and ensure these technologies benefit all. By approaching this frontier with wisdom, care, and a commitment to ethics, we can shape a future where the union of mind and machine enriches humanity and expands the horizons of possibility.

References

In this book, referencing is organized by chapter in order of use to facilitate a more streamlined and accessible reading experience. Each chapter includes its own set of references, allowing readers to easily locate sources relevant to the specific content discussed within that section. Readers are encouraged to consult the references at the end of each chapter for further exploration of the themes and concepts discussed.

Prologue & Introduction

1. Becher B. Brain-Computer Interfaces (BCI) Explained | Built In. builtin.com. Published July 25, 2023. https://builtin.com/hardware/brain-computer-interface-bci

2. George SC. 6 Superheroes of Ancient Civilizations. Discover Magazine. Published 2022. Accessed November 13, 2024. https://www.discovermagazine.com/planet-earth/6-superheroes-of-ancient-civilizations

3. Wikipedia Contributors. Artificial intelligence in fiction. Wikipedia. Published October 4, 2019. https://en.wikipedia.org/wiki/Artificial_intelligence_in_ficti on

4. Saha S, Mamun KA, Ahmed K, et al. Progress in Brain Computer Interface: Challenges and Opportunities. *Frontiers in Systems Neuroscience.* 2021;15. doi:https://doi.org/10.3389/fnsys.2021.578875

5. Musk E. An Integrated Brain-Machine Interface Platform With Thousands of Channels (Preprint). *Journal of Medical Internet Research.* 2019;21(10). doi:https://doi.org/10.2196/16194

6. The year of brain–computer interfaces. *Nature Electronics.* 2023;6(9):643-643. doi:https://doi.org/10.1038/s41928-023-01041-8

7. M Hainarosie, V Zainea, R Hainarosie. The evolution of cochlear implant technology and its clinical relevance. *Journal of Medicine and Life.* 2014;7(Spec Iss 2):1. https://pmc.ncbi.nlm.nih.gov/articles/PMC4391344/

8. Isa T, Fetz EE, Müller KR. Recent advances in brain–machine interfaces. *Neural Networks.* 2009;22(9):1201-1202. doi:https://doi.org/10.1016/j.neunet.2009.10.003

9. Mak JN, Wolpaw JR. Clinical Applications of Brain-Computer Interfaces: Current State and Future Prospects. *IEEE Reviews in Biomedical Engineering.* 2009;2:187-199. doi:https://doi.org/10.1109/RBME.2009.2035356

10. Isa T, Fetz EE, Müller KR. Recent advances in brain–machine interfaces. *Neural Networks.* 2009;22(9):1201-1202. doi:https://doi.org/10.1016/j.neunet.2009.10.003

11. Shih JJ, Krusienski DJ, Wolpaw JR. Brain-Computer Interfaces in Medicine. *Mayo Clinic Proceedings.* 2012;87(3):268-279. doi:https://doi.org/10.1016/j.mayocp.2011.12.008

12. Powers A. Mount Bonnell. Mount Bonnell. Published November 9, 2024. Accessed November 13, 2024.

https://www.mountbonnell.info/neural-nexus/the-role-of-ai-in-neuralinks-technology

13. Mann H. The future of brain-machine synchronisation. Polytechnique Insights. Published October 30, 2024. Accessed November 13, 2024. https://www.polytechnique-insights.com/en/columns/science/the-future-of-synchronising-brain-and-machine/

14. Why A. Neuro-Symbolic AI: Why Is It The Future of Artificial Intelligence. Startup Kitchen. Published May 20, 2024. Accessed November 13, 2024. https://startupkitchen.community/neuro-symbolic-ai-why-is-it-the-future-of-artificial-intelligence/

15. Pomeroy R. AI model uses human irrationality to predict our next moves. Big Think. Published June 12, 2024. Accessed November 13, 2024. https://bigthink.com/the-present/ai-model-decision-making/

16. Clanx.ai. Human-AI Collaboration: What it is and Why it Matters? Clanx.ai. Published 2023. https://clanx.ai/glossary/human-ai-colaboration

17. The Singularity Is Near: When Humans Transcend Biology: Kurzweil, Ray: 9780670033843: Amazon.com: Books. Amazon.com. Published 2024. Accessed November 13, 2024. https://www.amazon.com/Singularity-Near-Humans-Transcend-Biology/dp/0670033847

18. Livanis E, Voultsos P, Vadikolias K, Pantazakos P, Tsaroucha A. Understanding the Ethical Issues of Brain-Computer Interfaces (BCIs): A Blessing or the Beginning

of a Dystopian Future? *Cureus.* 2024;16(4). doi:https://doi.org/10.7759/cureus.58243

19. Haseltine WA. Brain-Machine Interfaces Spark Ethics Debate. 4 Areas Of Concern. *Forbes.* https://www.forbes.com/sites/williamhaseltine/2024/05/10/the-ethics-of-brain-machine-interfaces-concerns-and-considerations/. Published May 10, 2024.

20. Sun X, Ye B. The functional differentiation of brain–computer interfaces (BCIs) and its ethical implications. *Humanities and Social Sciences Communications.* 2023;10(1). doi:https://doi.org/10.1057/s41599-023-02419-x

21. Gil D. The Ethical Challenges of Connecting Our Brains to Computers. Scientific American. Published December 26, 2020. https://www.scientificamerican.com/article/the-ethical-challenges-of-connecting-our-brains-to-computers/

22. Sun X, Ye B. The functional differentiation of brain–computer interfaces (BCIs) and its ethical implications. *Humanities and Social Sciences Communications.* 2023;10(1). doi:https://doi.org/10.1057/s41599-023-02419-x

23. Valentin. Rethinking our consciousness: An approach to a scientifically feasible seamless mind-uploading. Research Features. Published January 13, 2023. https://researchfeatures.com/rethinking-consciousness/

24. on I. Mount Bonnell. Mount Bonnell. Published November 9, 2024. Accessed November 13, 2024. https://www.mountbonnell.info/neural-nexus/neuralinks-potential-impact-on-the-job-market-and-workforce-skills

25. Schenker JL. How Brain Computer Interface Technology Could Transform The Workplace - The Innovator. The Innovator. Published May 10, 2024. Accessed November 13, 2024. https://theinnovator.news/how-ai-powered-mind-reading-could-make-work-more-inclusive-and-productive/

26. U.S. Government Accountability Office. Science & Tech Spotlight: Brain-Computer Interfaces. www.gao.gov. Published September 8, 2022. https://www.gao.gov/products/gao-22-106118

27. Review TR. The Challenges of Regulating Brain-Machine Interfaces | The Regulatory Review. www.theregreview.org. Published November 24, 2022. https://www.theregreview.org/2022/11/24/fisher-the-challenges-of-regulating-brain-machine-interfaces/

Chapter 1

28. Willyard C. Paralyzed man walks naturally, thanks to wireless "bridge" between brain and spine. www.science.org. Published May 24, 2023. https://www.science.org/content/article/paralyzed-man-walks-naturally-thanks-wireless-bridge-brain-spine

29. Matt Angle. Google.com. Published 2019. Accessed November 13, 2024. https://scholar.google.com/citations?hl=en&user=jO1zRcQAAAAJ

30. Paradromics. Paradromics CEO Matt Angle on the Neuralink Demo. Paradromics.com. Published 2024. Accessed November 13, 2024.

https://www.paradromics.com/news/paradromics-ceo-matt-angle-on-the-neuralink-demo

31. Diamandis PH. Paradromics: Harness Your Brain's Full Power. Diamandis.com. Published 2021. Accessed November 13, 2024. https://www.diamandis.com/blog/harness-your-brains-full-power

32. CNBC. Inside Paradromics, the Neuralink competitor hoping to commercialize brain implants before the end of the decade. CNBC. Published June 20, 2024. Accessed November 13, 2024. https://www.cnbc.com/video/2024/06/20/paradromics-wants-to-make-brain-implants-a-reality-before-the-end-of-the-decade.html

33. Gomez M, Garratt C. Neuralink competitor Paradromics gears up to test its brain implant on humans. CNBC. Published June 21, 2024. Accessed November 13, 2024. https://www.cnbc.com/2024/06/21/paradromics-gears-up-to-test-its-brain-implant-on-humans.html

34. Neural Implant Podcast. Neural Implant podcast - the people behind Brain-Machine Interface revolutions: Jens Naumann on driving while blind using visual neural implants. neuralimplantpodcast.com. Published June 2017. https://neuralimplantpodcast.com/jens-naumann-on-driving-while-blind-using-visual-neural-implants

35. Balogh M. Man's high-tech paradise lost. thewhig. Published November 28, 2012. Accessed November 13,

2024. https://www.thewhig.com/2012/11/28/mans-high-tech-paradise-lost

36. Buckler G. Human-machine mergers promising, but reality yet to live up to hype. CBC. Published May 25, 2010. Accessed November 13, 2024. https://www.cbc.ca/news/science/human-machine-mergers-promising-but-reality-yet-to-live-up-to-hype-1.886289

37. KTVU FOX 2. Researchers use computer to give Bay Area man with ALS the gift of speech. KTVU FOX 2 San Francisco. Published August 15, 2024. Accessed November 13, 2024. https://www.ktvu.com/news/researchers-use-computer-give-bay-area-man-als-gift-speech

38. Mueller B. A.L.S. Stole His Voice. A.I. Retrieved It. *The New York Times.* https://www.nytimes.com/2024/08/14/health/als-ai-brain-implants.html. Published August 14, 2024.

39. Wade D. Researchers use artificial intelligence to help ALS patient speak again. Cbsnews.com. Published September 16, 2024. Accessed November 13, 2024. https://www.cbsnews.com/boston/news/als-patients-speech-ai-implants-braingate/

40. Wexler M. Brain-computer interface allows man with ALS to communicate. ALS News Today. Published August 19, 2024. Accessed November 13, 2024. https://alsnewstoday.com/news/brain-computer-interface-man-als-communicate/

41. Siliezar J. Brain-computer interface allows man with ALS to "speak" again. Brown University. Published October 29,

2024. https://www.brown.edu/news/2024-08-14/bci-speak-again

Chapter 2

42. Editorial. The history of Brain-Computer Interfaces (BCIs) - Timeline. Roboticsbiz.com. Published July 22, 2020. https://roboticsbiz.com/the-history-of-brain-computer-interfaces-bcis-timeline/#google_vignette

43. Kübler A. The history of BCI: From a vision for the future to real support for personhood in people with locked-in syndrome. *Neuroethics.* 2019;13. doi:https://doi.org/10.1007/s12152-019-09409-4

44. Mason E. Why was mummification used in Ancient Egypt, and why did they leave the heart in the body? History Extra. Published August 16, 2018. https://www.historyextra.com/period/ancient-egypt/why-egyptians-mummifed-dead-bodies-mummy-leave-heart-body/

45. Huffman C. Alcmaeon (Stanford Encyclopedia of Philosophy). Stanford.edu. Published 2017. https://plato.stanford.edu/entries/alcmaeon/

46. Hatfield G. René Descartes. Stanford Encyclopedia of Philosophy. Published December 3, 2008. https://plato.stanford.edu/entries/descartes/

47. Bern Dibner. Luigi Galvani | Italian physician and physicist. In: *Encyclopædia Britannica.* ; 2018. https://www.britannica.com/biography/Luigi-Galvani

48. Alessandro Volta. Britannica Kids. https://kids.britannica.com/students/article/Alessandro-Volta/339661

49. Dronkers NF, Plaisant O, Iba-Zizen MT, Cabanis EA. Paul Broca's historic cases: high resolution MR imaging of the brains of Leborgne and Lelong. *Brain*. 2007;130(5):1432-1441. doi:https://doi.org/10.1093/brain/awm042

50. Aphasia and praise of Pierre Paul Broca (1824–1880) | British Columbia Medical Journal. Bcmj.org. Published 2018. https://bcmj.org/blog/aphasia-and-praise-pierre-paul-broca-1824%E2%80%931880

51. Tizard B. THEORIES OF BRAIN LOCALIZATION FROM FLOURENS TO LASHLEY. *Medical History*. 1959;3(2):132-145. doi:https://doi.org/10.1017/s0025727300024418

52. Cohen AR. William Macewen and the first documented successful resection of a brain tumor. *Child's Nervous System: ChNS: Official Journal of the International Society for Pediatric Neurosurgery*. Published online January 17, 2023. doi:https://doi.org/10.1007/s00381-023-05825-3

53. Norbert Wiener | American mathematician | Britannica. In: *Encyclopædia Britannica*. ; 2019. https://www.britannica.com/biography/Norbert-Wiener

54. Max Planck Institute for Biological Cybernetics. From Cybernetics to AI: the pioneering work of Norbert Wiener - Max Planck Neuroscience. Max Planck Neuroscience -. Published April 25, 2024.

https://maxplanckneuroscience.org/from-cybernetics-to-ai-the-pioneering-work-of-norbert-wiener/

55. Neuroelectrics. Hans Berger: From Exploring Telepathy to Pioneering Electroencephalography. Neuroelectrics Blog - Latest news about EEG & Brain Stimulation. Published January 2, 2024. https://www.neuroelectrics.com/blog/2024/01/02/hans-berger-from-exploring-telepathy-to-pioneering-electroencephalography/

56. Delgado MB Fredrik Ekman, and José. Psychocivilization and Its Discontents: An Interview with José Delgado | Magnus Bärtås, Fredrik Ekman, and José Delgado. cabinetmagazine.org. https://www.cabinetmagazine.org/issues/2/bartas_ekman_delgado.php

57. Marzullo TC. The Missing Manuscript of Dr. Jose Delgado's Radio Controlled Bulls. *Journal of Undergraduate Neuroscience Education.* 2017;15(2):R29. https://pmc.ncbi.nlm.nih.gov/articles/PMC5480854/

58. Eshraghi AA, Nazarian R, Telischi FF, Rajguru SM, Truy E, Gupta C. The Cochlear Implant: Historical Aspects and Future Prospects. *The Anatomical Record: Advances in Integrative Anatomy and Evolutionary Biology.* 2012;295(11):1967-1980. doi:https://doi.org/10.1002/ar.22580

59. M Hainarosie, V Zainea, R Hainarosie. The evolution of cochlear implant technology and its clinical relevance.

Journal of Medicine and Life. 2014;7(Spec Iss 2):1.
https://pmc.ncbi.nlm.nih.gov/articles/PMC4391344/

60. Pope D. Neural Signals, Inc., Philip Kennedy, implanted
    brain microelectrode, ALS treatment, Emerge Medical.
    Neurotechreports.com. Published 2021.
    https://www.neurotechreports.com/pages/neuralsignalsprofi
    le.html

61. Georgia Tech Research Institute. Operating a computer
    with neural signals | GTRI Historical Archive. GTRI
    Historical Archive. Published 2023. Accessed November
    13, 2024.
    https://history.gtri.gatech.edu/innovations/operating-
    computer-neural-signals

62. Mak JN, Wolpaw JR. Clinical Applications of Brain-
    Computer Interfaces: Current State and Future Prospects.
    IEEE Reviews in Biomedical Engineering. 2009;2:187-
    199. doi:https://doi.org/10.1109/RBME.2009.2035356

63. Wander JD, Rao RP. Brain–computer interfaces: a
    powerful tool for scientific inquiry. *Current Opinion in
    Neurobiology.* 2014;25:70-75.
    doi:https://doi.org/10.1016/j.conb.2013.11.013

64. Orenstein D. People with paralysis control robotic arms
    using brain-computer interface. Brown.edu. Published
    2012. https://news.brown.edu/articles/2012/05/braingate2

65. Rejcek P. Robotic arms connected directly to brain of
    partially paralyzed man allows him to feed himself.
    Frontiers Science News. Published 2022.

https://www.frontiersin.org/news/2022/06/28/robotic-arms-feed-partially-paralyzed-man-bmi

66. NIH. Paralyzed individuals use thought-controlled robotic arm to reach and grasp. National Institutes of Health (NIH). Published August 31, 2015. Accessed November 13, 2024. https://www.nih.gov/news-events/news-releases/paralyzed-individuals-use-thought-controlled-robotic-arm-reach-grasp

Chapter 3

67. Rayi A, Murr N. Electroencephalogram. PubMed. Published 2021. https://www.ncbi.nlm.nih.gov/books/NBK563295/

68. Pilitsis JG. Deep Brain Stimulation. AANS. Published April 2024. https://www.aans.org/patients/conditions-treatments/deep-brain-stimulation/

69. Woodruff A. What is a Neuron? The University of Queensland. Published 2018. https://qbi.uq.edu.au/brain/brain-anatomy/what-neuron

70. Ludwig PE, Reddy V, Varacallo M. Neuroanatomy, Neurons. PubMed. Published July 24, 2023. https://www.ncbi.nlm.nih.gov/books/NBK441977/

71. Associates N. Brainwave Frequencies: What Are They? | NHA Blog. NeuroHealth Associates. Published January 11, 2022. https://nhahealth.com/brainwave-frequencies-what-are-they/

72. Lazarou I, Nikolopoulos S, Petrantonakis PC, Kompatsiaris I, Tsolaki M. EEG-Based Brain–Computer Interfaces for

Communication and Rehabilitation of People with Motor Impairment: A Novel Approach of the 21st Century. *Frontiers in Human Neuroscience.* 2018;12. doi:https://doi.org/10.3389/fnhum.2018.00014

73. Villines Z. Neurofeedback for ADHD: Does it work? What to expect. www.medicalnewstoday.com. Published October 2018. https://www.medicalnewstoday.com/articles/315261

74. Hafeez T, Umar Saeed SM, Arsalan A, Anwar SM, Ashraf MU, Alsubhi K. EEG in game user analysis: A framework for expertise classification during gameplay. Fernández-Hilario A, ed. *PLOS ONE.* 2021;16(6):e0246913. doi:https://doi.org/10.1371/journal.pone.0246913

75. Pacia SV. Sub-Scalp Implantable Telemetric EEG (SITE) for the Management of Neurological and Behavioral Disorders beyond Epilepsy. *Brain Sciences.* 2023;13(8):1176-1176. doi:https://doi.org/10.3390/brainsci13081176

76. University of Florida. Subdural Electrode Monitoring» Department of Neurology» College of Medicine» University of Florida. Ufl.edu. Published 2024. Accessed November 13, 2024. https://neurology.ufl.edu/divisions/epilepsy/epilepsy-surgery-program/subdural-electrode-monitoring/

77. Zhao ZP, Nie C, Jiang CT, et al. Modulating Brain Activity with Invasive Brain–Computer Interface: A Narrative Review. *Brain Sciences.* 2023;13(1):134. doi:https://doi.org/10.3390/brainsci13010134

78. Liu K, Zhang H, Hu M, et al. The Past, Present, and Future of In Vivo-Implantable Recording Microelectrodes: The Neural Interfaces. *Materials Advances.* 2024;5(12):4958-4973. doi:https://doi.org/10.1039/d3ma01105d

79. Simeral JD, Kim S-P, Black MJ, Donoghue JP, Hochberg LR. Neural control of cursor trajectory and click by a human with tetraplegia 1000 days after implant of an intracortical microelectrode array. *Journal of Neural Engineering.* 2011;8(2):025027. doi:https://doi.org/10.1088/1741-2560/8/2/025027

80. BrainGate. About Braingate. BrainGate. Published December 4, 2015. https://www.braingate.org/about-braingate/

81. Freudenburg ZV, Branco MP, Leinders S, et al. Sensorimotor ECoG Signal Features for BCI Control: A Comparison Between People With Locked-In Syndrome and Able-Bodied Controls. *Frontiers in Neuroscience.* 2019;13. doi:https://doi.org/10.3389/fnins.2019.01058

82. Caldwell DJ, Ojemann JG, Rao RPN. Direct Electrical Stimulation in Electrocorticographic Brain–Computer Interfaces: Enabling Technologies for Input to Cortex. *Frontiers in Neuroscience.* 2019;13. doi:https://doi.org/10.3389/fnins.2019.00804

83. Liu X, Gong Y, Jiang Z, Stevens T, Li W. Flexible high-density microelectrode arrays for closed-loop brain–machine interfaces: a review. *Frontiers in Neuroscience.* 2024;18. doi:https://doi.org/10.3389/fnins.2024.1348434

84. Neurolink. Neuralink. neuralink.com. Published 2024. https://neuralink.com/

85. Capitol Technology University. Neuralink's Brain Chip: How It Works and What It Means | Capitol Technology University. www.captechu.edu. Published February 9, 2024. https://www.captechu.edu/blog/neuralinks-brain-chip-how-it-works-and-what-it-means

86. Purvish Mahendra Parikh, Ajit Venniyoor. Neuralink and Brain–Computer Interface—Exciting Times for Artificial Intelligence. *South Asian journal of cancer.* Published online April 15, 2024. doi:https://doi.org/10.1055/s-0043-1774729

87. Ma Q, Gao W, Xiao Q, et al. Directly wireless communication of human minds via non-invasive brain-computer-metasurface platform. *eLight.* 2022;2(1). doi:https://doi.org/10.1186/s43593-022-00019-x

88. International Journal of Molecular Sciences. International Journal of Molecular Sciences. Mdpi.com. Published 2024. Accessed November 13, 2024. https://www.mdpi.com/journal/ijms/special_issues/biocompa_materials

89. University of Manchester. Applications - Graphene - The University of Manchester. Manchester.ac.uk. Published 2019. https://www.graphene.manchester.ac.uk/learn/applications/

Chapter 4

90. Zhang X, Ma Z, Zheng H, et al. The combination of brain-computer interfaces and artificial intelligence: applications and challenges. *Annals of Translational Medicine.* 2020;8(11). doi:https://doi.org/10.21037/atm.2019.11.109

91. UCL. AI used to decode brain signals and predict behaviour. UCL News. Published August 17, 2021. https://www.ucl.ac.uk/news/2021/aug/ai-used-decode-brain-signals-and-predict-behaviour

92. Ruberg S, Ward J. From brain waves, this AI can sketch what you're picturing. NBC News. Published March 2023. https://www.nbcnews.com/tech/tech-news/brain-waves-ai-can-sketch-picturing-rcna76096

93. Girdler B, Caldbeck W, Bae J. Neural Decoders Using Reinforcement Learning in Brain Machine Interfaces: A Technical Review. *Frontiers in Systems Neuroscience.* 2022;16. doi:https://doi.org/10.3389/fnsys.2022.836778

94. Mathias Vukelić, Bui M, Vorreuther A, Lingelbach K. Combining brain-computer interfaces with deep reinforcement learning for robot training: a feasibility study in a simulation environment. *Frontiers in neuroergonomics.* 2023;4. doi:https://doi.org/10.3389/fnrgo.2023.1274730

95. Rosso C. Brain-Computer Interfaces Boosted by Novel AI Algorithm. Psychology Today. Published 2024. Accessed November 13, 2024. https://www.psychologytoday.com/us/blog/the-future-

brain/202409/brain-computer-interfaces-boosted-by-novel-ai-algorithm

96. Mathias Vukelić, Bui M, Vorreuther A, Lingelbach K. Combining brain-computer interfaces with deep reinforcement learning for robot training: a feasibility study in a simulation environment. *Frontiers in neuroergonomics.* 2023;4. doi:https://doi.org/10.3389/fnrgo.2023.1274730

97. Ahn M, Jun SC, Yeom HG, Cho H. Editorial: Deep Learning in Brain-Computer Interface. *Frontiers in Human Neuroscience.* 2022;16. doi:https://doi.org/10.3389/fnhum.2022.927567

98. McMillan T, McMillan T. Breakthrough in Brain-Computer Interfaces: Scientists Use AI Neural Decoding to Predict Mouse Movements with 95% Accuracy - The Debrief. The Debrief. Published March 27, 2024. Accessed November 13, 2024. https://thedebrief.org/breakthrough-in-brain-computer-interfaces-scientists-use-ai-neural-decoding-to-predict-mouse-movements-with-95-accuracy/

99. Moore T. AI is the key to astonishing breakthrough that allowed paralysed man to walk again. Sky News. Published May 2023. https://news.sky.com/story/ai-is-the-key-to-astonishing-breakthrough-that-allowed-paralysed-man-to-walk-again-12888291

100. Rosso C. How Synergistic Is the Combo of AI and Humans? Psychology Today. Published 2024. Accessed November 13, 2024. https://www.psychologytoday.com/us/blog/the-future-

brain/202411/how-synergistic-is-the-combo-of-ai-and-humans

101. Salk Institute. Teaching artificial intelligence to adapt. ScienceDaily. Published 2020. Accessed November 13, 2024. https://www.sciencedaily.com/releases/2020/12/201216155 201.htm

102. University of Surrey. Using AI-related technologies can significantly enhance human cognition, finds new study | University of Surrey. www.surrey.ac.uk. Published December 2023. https://www.surrey.ac.uk/news/using-ai-related-technologies-can-significantly-enhance-human-cognition-finds-new-study

103. Nyholm S. Artificial Intelligence and Human Enhancement: Can AI Technologies Make Us More (Artificially) Intelligent? *Cambridge Quarterly of Healthcare Ethics.* Published online August 30, 2023:1-13. doi:https://doi.org/10.1017/S0963180123000464

104. Jiang L, Stocco A, Losey DM, Abernethy JA, Prat CS, Rao RPN. BrainNet: A Multi-Person Brain-to-Brain Interface for Direct Collaboration Between Brains. *Scientific Reports.* 2019;9(1). doi:https://doi.org/10.1038/s41598-019-41895-7

105. Moreno-Calderón S, Víctor Martínez-Cagigal, Santamaría-Vázquez E, Pérez-Velasco S, Marcos-Martínez D, Hornero R. Combining brain-computer interfaces and multiplayer video games: an application based on c-VEPs.

Frontiers in Human Neuroscience. 2023;17. doi:https://doi.org/10.3389/fnhum.2023.1227727

Chapter 5

106. Oregon Health and Science University Foundation. The future is now: Albert Chi, MD, and mind-controlled prosthetics. OHSU Foundation. Published May 10, 2018. https://ohsufoundation.org/stories/the-future-is-now-albert-chi-md-and-mind-controlled-prosthetics/

107. News C. Pioneer of world's most advanced prosthetic arm has Tucson ties. Q2 News (KTVQ). Published June 28, 2019. Accessed November 13, 2024. https://www.ktvq.com/cnn-regional/2019/06/28/pioneer-of-worlds-most-advanced-prosthetic-arm-has-tucson-ties/

108. Johns Hopkins. Amputee Makes Music with the Modular Prosthetic Limb | Johns Hopkins University Applied Physics Laboratory. www.jhuapl.edu. Published March 2021. https://www.jhuapl.edu/news/news-releases/210318-home-study-Modular-Prosthetic-Limb-Matheny-piano-Amazing-Grace

109. Taska. How does a myoelectric hand work? Taskaprosthetics.com. Published 2024. https://www.taskaprosthetics.com/news/myoelectric-hand

110. McManus L, De Vito G, Lowery MM. Analysis and Biophysics of Surface EMG for Physiotherapists and Kinesiologists: toward a Common Language with Rehabilitation Engineers. *Frontiers in Neurology.* 2020;11(1). doi:https://doi.org/10.3389/fneur.2020.576729

111. The Ohio State University. Targeted Muscle Reinnervation I Ohio State Medical Center. wexnermedical.osu.edu. https://wexnermedical.osu.edu/plastic-surgery/restorative-surgery-and-repair/targeted-muscle-reinnervation

112. Richman M. Study probes user satisfaction with upper-limb prostheses. www.research.va.gov. Published February 2020. https://www.research.va.gov/currents/0220-Study-probes-user-satisfaction-with-upper-limb-prostheses.cfm

113. Pezzin LE, Dillingham TR, MacKenzie EJ, Ephraim P, Rossbach P. Use and satisfaction with prosthetic limb devices and related services 11No commercial party having a direct financial interest in the results of the research supporting this article has or will confer a benefit on the author(s) or on any organization with which the author(s) is/are associated. *Archives of Physical Medicine and Rehabilitation.* 2004;85(5):723-729. doi:https://doi.org/10.1016/j.apmr.2003.06.002

114. Chalmers University of Technology. Breakthrough paves the way for next generation of vision implants. ScienceDaily. Published 2024. Accessed November 13, 2024. https://www.sciencedaily.com/releases/2024/05/240507145945.htm

115. NIH. Personalized deep brain stimulation for Parkinson's disease. National Institutes of Health (NIH). Published September 9, 2024. https://www.nih.gov/news-events/nih-

research-matters/personalized-deep-brain-stimulation-parkinson-s-disease

116. John Hopkins Medicine. Deep Brain Stimulation. John Hopkins Medicine. Published 2019. https://www.hopkinsmedicine.org/health/treatment-tests-and-therapies/deep-brain-stimulation

117. Little S, Beudel M, Zrinzo L, et al. Bilateral adaptive deep brain stimulation is effective in Parkinson's disease . Bmj.com. Published 2014. Accessed November 13, 2024. https://jnnp.bmj.com/content/87/7/717

118. Craine T. Deep brain stimulation surgery helps Parkinson's patient get her life back to normal. University of Iowa Health Care. Published 2016. Accessed November 13, 2024. https://uihc.org/patient-story/deep-brain-stimulation-surgery-helps-parkinsons-patient-get-her-life-back-normal

119. Hohl K, Giffhorn M, Jackson S, Jayaraman A. A framework for clinical utilization of robotic exoskeletons in rehabilitation. *Journal of NeuroEngineering and Rehabilitation*. 2022;19(1). doi:https://doi.org/10.1186/s12984-022-01083-7

120. Streed J. How spinal stimulation research is working to restore function after paralysis. Mayo Clinic News Network. Published June 13, 2019. https://newsnetwork.mayoclinic.org/discussion/alumni-how-spinal-stimulation-research-is-working-to-restore-function-after-paralysis/

121. Madson R. Spinal cord stimulation, physical therapy help paralyzed man stand, walk with assistance. Mayo Clinic News Network. Published September 24, 2018. https://newsnetwork.mayoclinic.org/discussion/spinal-cord-stimulation-physical-therapy-help-paralyzed-man-stand-walk-with-assistance/

122. TMSi. What are the Applications of HD-EMG? Tmsi.com. Published January 8, 2024. https://info.tmsi.com/blog/what-are-the-applications-of-hd-emg

123. Malešević N, Olsson A, Sager P, et al. A database of high-density surface electromyogram signals comprising 65 isometric hand gestures. *Scientific Data*. 2021;8(1). doi:https://doi.org/10.1038/s41597-021-00843-9

124. Kuruganti U, Pradhan A, Toner J. High-Density Electromyography Provides Improved Understanding of Muscle Function for Those With Amputation. *Frontiers in Medical Technology*. 2021;3:690285. doi:https://doi.org/10.3389/fmedt.2021.690285

125. Archives of Orthopaedics. EMG Signal Processing for Hand Motion Pattern Recognition Using Machine Learning Algorithms. *Archives of Orthopaedics*. 2020;(1). doi:https://doi.org/10.33696/orthopaedics.1.005

126. Reaz MBI, Hussain MS, Mohd-Yasin F. Techniques of EMG signal analysis: detection, processing, classification and applications. *Biological Procedures Online*. 2006;8(1):11-35. doi:https://doi.org/10.1251/bpo115

127. Resnik L, Huang H (Helen), Winslow A, Crouch DL, Zhang F, Wolk N. Evaluation of EMG pattern recognition for upper limb prosthesis control: a case study in comparison with direct myoelectric control. *Journal of NeuroEngineering and Rehabilitation*. 2018;15(1). doi:https://doi.org/10.1186/s12984-018-0361-3

128. Blankertz B, Tangermann M, Vidaurre C, et al. The Berlin Brain–Computer Interface: Non-Medical Uses of BCI Technology. *Frontiers in Neuroscience*. 2010;4. doi:https://doi.org/10.3389/fnins.2010.00198

129. Nijholt A, Contreras-Vidal JL, Jeunet C, Aleksander Väljamäe. Editorial: Brain-Computer Interfaces for Non-clinical (Home, Sports, Art, Entertainment, Education, Well-Being) Applications. *Frontiers in computer science*. 2022;4. doi:https://doi.org/10.3389/fcomp.2022.860619

130. BCI Games. BCI Games | Home for Brain Computing Interface games and development. bci.games. https://bci.games/

131. Gajbhiye S. New brain-computer interface lets users game with their minds. Earth.com. Published 2024. Accessed November 13, 2024. https://www.earth.com/news/new-brain-computer-interface-lets-users-game-with-their-minds/

132. Douibi K, Le Bars S, Lemontey A, Nag L, Balp R, Breda G. Toward EEG-Based BCI Applications for Industry 4.0: Challenges and Possible Applications. *Frontiers in Human Neuroscience*. 2021;15. doi:https://doi.org/10.3389/fnhum.2021.705064

133. Gonfalonieri A. What Brain-Computer Interfaces Could Mean for the Future of Work. Harvard Business Review. Published October 6, 2020. https://hbr.org/2020/10/what-brain-computer-interfaces-could-mean-for-the-future-of-work

134. Anderson J, Rainie L, Vogels E. Experts Say the "New Normal" in 2025 Will Be Far More Tech-Driven, Presenting More Big Challenges. Pew Research Center. Published February 18, 2021. https://www.pewresearch.org/internet/2021/02/18/experts-say-the-new-normal-in-2025-will-be-far-more-tech-driven-presenting-more-big-challenges/

Chapter 6

135. Maiseli B, Abdalla AT, Massawe LV, et al. Brain–computer interface: trend, challenges, and threats. *Brain Informatics*. 2023;10(1):20. doi:https://doi.org/10.1186/s40708-023-00199-3

136. Salahuddin U, Gao PX. Signal Generation, Acquisition, and Processing in Brain Machine Interfaces: A Unified Review. *Frontiers in Neuroscience*. 2021;15:728178. doi:https://doi.org/10.3389/fnins.2021.728178

137. Oweiss KG, Badreldin IS. Neuroplasticity subserving the operation of brain–machine interfaces. *Neurobiology of Disease*. 2015;83:161-171. doi:https://doi.org/10.1016/j.nbd.2015.05.001

138. Rebsamen B, Guan C, Zhang H, et al. A Brain Controlled Wheelchair to Navigate in Familiar Environments. *IEEE*

Transactions on Neural Systems and Rehabilitation Engineering. 2010;18(6):590-598. doi:https://doi.org/10.1109/tnsre.2010.2049862

139. Saha S, Mamun KA, Ahmed K, et al. Progress in Brain Computer Interface: Challenges and Opportunities. *Frontiers in Systems Neuroscience.* 2021;15. doi:https://doi.org/10.3389/fnsys.2021.578875

140. Hello Future. The challenge: Brain-computer interfaces that work for everyone - Hello Future Orange. Hello Future. Published January 16, 2024. https://hellofuture.orange.com/en/the-challenge-brain-computer-interfaces-that-work-for-everyone/

141. Liu Z, Tang J, Gao B, et al. Neural signal analysis with memristor arrays towards high-efficiency brain–machine interfaces. *Nature Communications.* 2020;11(1). doi:https://doi.org/10.1038/s41467-020-18105-4

142. Robinson N, Chouhan T, Mihelj E, et al. Design Considerations for Long Term Non-invasive Brain Computer Interface Training With Tetraplegic CYBATHLON Pilot. *Frontiers in Human Neuroscience.* 2021;15. doi:https://doi.org/10.3389/fnhum.2021.648275

143. Markets and Markets. Brain Computer Interface Market Size, Share & Trends [2029]. MarketsandMarkets. https://www.marketsandmarkets.com/Market-Reports/brain-computer-interface-market-64821525.html

144. Haseltine WA. Current Costs And Technology Limit Brain-Machine Interfaces. Forbes. Published September 22, 2023.

https://www.forbes.com/sites/williamhaseltine/2023/09/22/
current-costs-and-technology-limit-brain-machine-
interfaces/

145. McCrimmon CM, Fu JL, Wang M, et al. Performance
Assessment of a Custom, Portable, and Low-Cost Brain–
Computer Interface Platform. *IEEE Transactions on
Biomedical Engineering.* 2017;64(10):2313-2320.
doi:https://doi.org/10.1109/tbme.2017.2667579

146. Neuralink's in. Mount Bonnell. Mount Bonnell.
Published November 9, 2024. Accessed November 13,
2024. https://www.mountbonnell.info/neural-
nexus/challenges-in-scaling-up-neuralink-production

147. Shih JJ, Krusienski DJ, Wolpaw JR. Brain-Computer
Interfaces in Medicine. *Mayo Clinic Proceedings.*
2012;87(3):268-279.
doi:https://doi.org/10.1016/j.mayocp.2011.12.008

148. American Psychological Association. Speaking of
Psychology: The future of brain-computer interfaces, with
Nicholas Hatsopoulos, PhD. Apa.org. Published 2024.
Accessed November 13, 2024.
https://www.apa.org/news/podcasts/speaking-of-
psychology/brain-computer-interfaces

149. Ayala G, Haslacher D, Krol LR, Soekadar SR, Zander TO.
Assessment of mental workload across cognitive tasks using
a passive brain-computer interface based on mean
negative theta-band amplitudes. *Frontiers in
neuroergonomics.* 2023;4.
doi:https://doi.org/10.3389/fnrgo.2023.1233722

150. Caspar EA, De Beir A, Lauwers G, Cleeremans A, Vanderborght B. How using brain-machine interfaces influences the human sense of agency. Aspell JE, ed. *PLOS ONE*. 2021;16(1):e0245191. doi:https://doi.org/10.1371/journal.pone.0245191

151. Davidoff EJ. Agency and Accountability: Ethical Considerations for Brain-Computer Interfaces. *The Rutgers journal of bioethics*. 2020;11:9. https://pmc.ncbi.nlm.nih.gov/articles/PMC7654969/

152. Silva GA. A New Frontier: The Convergence of Nanotechnology, Brain Machine Interfaces, and Artificial Intelligence. *Frontiers in Neuroscience*. 2018;12. doi:https://doi.org/10.3389/fnins.2018.00843

153. Zhang X, Ma Z, Zheng H, et al. The combination of brain-computer interfaces and artificial intelligence: applications and challenges. *Annals of Translational Medicine*. 2020;8(11). doi:https://doi.org/10.21037/atm.2019.11.109

154. Opris I, Lebedev MA, Pulgar VM, Vidu R, Enachescu M, Casanova MF. Editorial: Nanotechnologies in Neuroscience and Neuroengineering. *Frontiers in Neuroscience*. 2020;14. doi:https://doi.org/10.3389/fnins.2020.00033

155. Glannon W. Ethical issues with brain-computer interfaces. *Frontiers in Systems Neuroscience*. 2014;8. doi:https://doi.org/10.3389/fnsys.2014.00136

Chapter 7

156. Tseng J, Poppenk J. Brain meta-state transitions demarcate thoughts across task contexts exposing the mental noise of trait neuroticism. *Nature Communications.* 2020;11(1):3480. doi:https://doi.org/10.1038/s41467-020-17255-9

157. Sanders L. Can privacy coexist with technology that reads and changes brain activity? Science News. Published February 11, 2021. https://www.sciencenews.org/article/technology-brain-activity-read-change-thoughts-privacy-ethics

158. Manar Alohaly. Brain computer interface is growing but what are the risks? World Economic Forum. Published June 14, 2024. https://www.weforum.org/stories/2024/06/the-brain-computer-interface-market-is-growing-but-what-are-the-risks/

159. Natural-born Cyborgs. Google Books. Published 2024. Accessed November 14, 2024. https://books.google.de/books/about/Natural_born_Cyborgs.html?id=8JXaK3sREXQC&redir_esc=y

160. Dubljević V, Coin A, Shipman M. Studies Outline Key Ethical Questions Surrounding Brain-Computer Interface Tech. NC State News. Published November 10, 2020. https://news.ncsu.edu/2020/11/brain-computer-interface-ethics/

161. Sankaran N, Moses D, Chiong W, Chang EF. Recommendations for promoting user agency in the

design of speech neuroprostheses. *Frontiers in human neuroscience*. 2023;17. doi:https://doi.org/10.3389/fnhum.2023.1298129

162. Caspar EA, De Beir A, Lauwers G, Cleeremans A, Vanderborght B. How using brain-machine interfaces influences the human sense of agency. Aspell JE, ed. *PLOS ONE*. 2021;16(1):e0245191. doi:https://doi.org/10.1371/journal.pone.0245191

163. Carmena JM, Millán J del R. Brain Machine Interfaces: Your Brain in Action. *Frontiers for Young Minds*. 2013;1. doi:https://doi.org/10.3389/frym.2013.00007

164. Haseltine WA. The Need for Ethical Regulation of Brain-Machine Interface Technologies. Inside Precision Medicine. Published August 2024. https://www.insideprecisionmedicine.com/topics/translatio nal-research/the-need-for-ethical-regulation-of-brain-machine-interface-technologies/

165. Dickey J. International Association of Privacy Professionals. iapp.org. Published June 11, 2024. https://iapp.org/news/a/navigating-the-legal-and-ethical-landscape-of-brain-computer-interfaces-insights-from-colorado-and-minnesota

166. Xia R, Yang S. Factors influencing the social acceptance of brain-computer interface technology among Chinese general public: an exploratory study. *Frontiers in Human Neuroscience*. 2024;18. doi:https://doi.org/10.3389/fnhum.2024.1423382

167. Lavazza A. Can Neuromodulation also Enhance Social Inequality? Some Possible Indirect Interventions of the State. *Frontiers in Human Neuroscience.* 2017;11. doi:https://doi.org/10.3389/fnhum.2017.00113

168. Sun X, Ye B. The functional differentiation of brain–computer interfaces (BCIs) and its ethical implications. *Humanities and Social Sciences Communications.* 2023;10(1). doi:https://doi.org/10.1057/s41599-023-02419-x

169. Gordon EC, Seth AK. Ethical considerations for the use of brain–computer interfaces for cognitive enhancement. *PLoS Biology.* 2024;22(10):e3002899-e3002899. doi:https://doi.org/10.1371/journal.pbio.3002899

170. OECD Science, Technology and Industry Working Papers. Brain-computer interfaces and the governance system. OECD. Published 2024. Accessed November 14, 2024. https://www.oecd.org/en/publications/brain-computer-interfaces-and-the-governance-system_18d86753-en.html

Chapter 8

171. Liu Z, Tang J, Gao B, et al. Neural signal analysis with memristor arrays towards high-efficiency brain–machine interfaces. *Nature Communications.* 2020;11(1). doi:https://doi.org/10.1038/s41467-020-18105-4

172. Valencia D, Leone G, Keller N, Mercier PP, Alimohammad A. Power-efficient in vivo brain-machine interfaces via brain-state estimation. *Journal of Neural*

Engineering. 2023;20(1):016032.
doi:https://doi.org/10.1088/1741-2552/acb385

173. The Galli Group. The Galli Group. Uchicago.edu.
Published 2024. Accessed November 14, 2024.
https://galligroup.uchicago.edu/People/galli.php

174. The Galli Group. Materials for Neuromorphic Devices.
Uchicago.edu. Published 2022. Accessed November 14,
2024.
https://galligroup.uchicago.edu/Research/neuromorphic.p
hp

175. Caballar R, Stryker C. What Is Neuromorphic
Computing? | IBM. www.ibm.com. Published June 27,
2024. https://www.ibm.com/think/topics/neuromorphic-
computing

176. Kim MK, Park Y, Kim IJ, Lee JS. Emerging Materials for
Neuromorphic Devices and Systems. *iScience.*
2020;23(12):101846.
doi:https://doi.org/10.1016/j.isci.2020.101846

177. Hoffmann A, Ramanathan S, Grollier J, et al. Quantum
materials for energy-efficient neuromorphic computing:
Opportunities and challenges. *APL Materials.* 2022;10(7).
doi:https://doi.org/10.1063/5.0094205

178. Saïghi S, Mayr CG, Serrano-Gotarredona T, et al.
Plasticity in memristive devices for spiking neural
networks. *Frontiers in Neuroscience.* 2015;9.
doi:https://doi.org/10.3389/fnins.2015.00051

179. Huang HM, Wang Z, Wang T, Xiao Y, Guo X. Artificial Neural Networks Based on Memristive Devices: From Device to System. *Advanced Intelligent Systems*. 2020;2(12):2000149. doi:https://doi.org/10.1002/aisy.202000149

180. Wang J, Zhu Y, Zhu L, Chen C, Wan Q. Emerging Memristive Devices for Brain-Inspired Computing and Artificial Perception. *Frontiers in Nanotechnology*. 2022;4. doi:https://doi.org/10.3389/fnano.2022.940825

181. Wang L, Lu SR, Wen J. Recent Advances on Neuromorphic Systems Using Phase-Change Materials. *Nanoscale Research Letters*. 2017;12(1). doi:https://doi.org/10.1186/s11671-017-2114-9

182. Pecqueur S, Vuillaume D, Alibart F. Perspective: Organic electronic materials and devices for neuromorphic engineering featured. Aip.org. Published 2024. Accessed November 14, 2024. https://pubs.aip.org/aip/jap/article/124/15/151902/348082/Perspective-Organic-electronic-materials-and

183. Imke Krauhausen, Griggs S, McCulloch I, Jaap, Paschalis Gkoupidenis, van. Bio-inspired multimodal learning with organic neuromorphic electronics for behavioral conditioning in robotics. *Nature Communications*. 2024;15(1). doi:https://doi.org/10.1038/s41467-024-48881-2

184. Max Planck Institute for Polymer Research. Organic neuromorphic devices. Mpip-mainz.mpg.de. Published

2024. Accessed November 14, 2024. https://sites.mpip-mainz.mpg.de/93239/Organic-neuromorphic-devices

185. Boi F, Moraitis T, De Feo V, et al. A Bidirectional Brain-Machine Interface Featuring a Neuromorphic Hardware Decoder. *Frontiers in Neuroscience*. 2016;10. doi:https://doi.org/10.3389/fnins.2016.00563

186. Qi Y, Chen J, Wang Y. Neuromorphic computing facilitates deep brain-machine fusion for high-performance neuroprosthesis. *Frontiers in Neuroscience*. 2023;17. doi:https://doi.org/10.3389/fnins.2023.1153985

187. Gumyusenge A, Melianas A, Keene ST, Salleo A. Materials Strategies for Organic Neuromorphic Devices. *Annual Review of Materials Research*. 2021;51(1):47-71. doi:https://doi.org/10.1146/annurev-matsci-080619-111402

188. Technology Networks. AI Algorithm Separates Brain Patterns Related to Particular Behaviors. Neuroscience from Technology Networks. Published September 10, 2024. Accessed November 14, 2024. https://www.technologynetworks.com/neuroscience/news/ai-algorithm-separates-brain-patterns-related-to-particular-behaviors-390749

189. University of Southern California. New AI can ID brain patterns related to specific behavior. ScienceDaily. Published 2024. https://www.sciencedaily.com/releases/2024/09/240909175239.htm

190. Miles S. Unlocking the mind: Why and how AI can help decode brain activity. Electronicspecifier.com. Published

2024. Accessed November 14, 2024.
https://www.electronicspecifier.com/products/artificial-intelligence/unlocking-the-mind-why-and-how-ai-can-help-decode-brain-activity

191. Nanobaly. Understanding pattern recognition in AI. en.innovatiana.com. Published 2024. https://en.innovatiana.com/post/pattern-recognition-in-ai

192. Qi Y, Chen J, Wang Y. Neuromorphic computing facilitates deep brain-machine fusion for high-performance neuroprosthesis. *Frontiers in Neuroscience.* 2023;17. doi:https://doi.org/10.3389/fnins.2023.1153985

193. Merk T, Peterson V, Köhler R, Haufe S, Richardson RM, Neumann WJ. Machine learning based brain signal decoding for intelligent adaptive deep brain stimulation. *Experimental Neurology.* 2022;351:113993. doi:https://doi.org/10.1016/j.expneurol.2022.113993

194. Valeriani D, Santoro F, Ienca M. The present and future of neural interfaces. *Frontiers in Neurorobotics.* 2022;16. doi:https://doi.org/10.3389/fnbot.2022.953968

195. USC Viterbi Staff. New AI Algorithm Enables Advanced Real-Time Decoding for Neurotechnologies - USC Viterbi | School of Engineering. USC Viterbi | School of Engineering. Published May 16, 2024. Accessed November 14, 2024. https://viterbischool.usc.edu/news/2023/12/new-ai-algorithm-enables-advanced-real-time-decoding-for-neurotechnologies/

196. Drug Target Review. Novel AI algorithm improves decoding accuracy of brain signals. Drug Target Review. Published December 15, 2023. Accessed November 14, 2024. https://www.drugtargetreview.com/news/113183/novel-ai-algorithm-improves-decoding-accuracy-of-brain-signals/

197. Schollmeier R. Unlocking Autonomy with Brain-Computer and Brain-Machine Interfaces: Current Insights and Future Frontiers. ncs_hub_dnn091100_dev. Published August 7, 2024. https://currents.neurocriticalcare.org/Leading-Insights/Article/unlocking-autonomy-with-brain-computer-and-brain-machine-interfaces-current-insights-and-future-frontiers

198. Lebedev MA, Nicolelis MAL. Brain-Machine Interfaces: From Basic Science to Neuroprostheses and Neurorehabilitation. *Physiological Reviews*. 2017;97(2):767-837. doi:https://doi.org/10.1152/physrev.00027.2016

Chapter 9

199. William Gibson. *Neuromancer by William Gibson.*; 1984. https://archive.org/details/neuromancer_202209

200. *The Matrix*. Warner Bros; 1999.

201. Brooker, C. Black Mirror. Published online 2011.

202. Shirow, M. Ghost in the Shell [Manga].

www.ingramcontent.com/pod-product-compliance
Lightning Source LLC
Chambersburg PA
CBHW070930210326
41520CB00021B/6875